COLLINS

A-Z OF GARDEN
PESTS & PROBLEMS

COLLINS

A-Z OF GARDEN PESTS & PROBLEMS

Ian G. Walls
N.D.H., C.D.H., A.Inst.P.R.A.

Collins
Glasgow and London

First published 1979
Published by William Collins Sons and Company Ltd
Glasgow and London

© Ian G. Walls

Line drawings by Dr Dietrich L. Burkel

© 1979 William Collins Sons and Company Ltd

ISBN 0 00 410401 3

Printed in Great Britain

Contents

List of colour illustrations and acknowledgements

Introduction

This book deals with pests, diseases and other problems which affect garden, greenhouse and commercial glasshouse plants, their symptoms and their control. A plant, bush or tree causes concern when its appearance to our eyes is not 'normal' in some respects. Leaves, stems, flowers or fruit can show symptoms of discolouration, abnormal size, fungal or bacterial infection, pest attack, wilting, rotting or poor growth generally.

Damage may be severe and readily noticed or very slight and only discernible to the fastidious gardener. When something is wrong or not quite up to standard, the gardener should try to find out why this should be, so that palliative or remedial measures can be taken. Even where things have reached a stage when nothing can be done for the plant or crop, steps must be taken to ensure that the same thing does not happen to adjacent plants or plants in the following season.

The First Inspection

It is usually visual inspection with the naked eye which first detects that all is not well with a plant. Closer inspection, perhaps with a magnifying glass and careful comparison with other plants in other gardens, usually determines whether the plant is ailing. The possible cause may then be sought in the appropriate plant or pest entry in the pages of this book. However, the commercial grower should probably also call in his adviser, especially when large scale cropping is involved. While he may have his own ideas about what is wrong, a second opinion is always valuable, especially where a livelihood is involved.

A second opinion can sometimes be obtained from any local garden centre, garden supplier or nurseryman. The trouble may be simple and easily solved, especially by an experienced nurseryman. Garden centres usually have a fair idea of local problems as there tend to be attacks of pests and diseases peculiar to a district in a particular season. If, however, things are not straightforward, further expert help will need to be sought.

Horticultural Advice

Advisory Services are operated by horticultural bodies, for example the Royal Horticultural Society, Parks Department, local

Education Authority, University or Agricultural College. The availability of advice varies in different parts of the country. For England and Wales, there is the Agricultural Development Advisory Service (ADAS); for Scotland, there are the Agricultural Colleges; for Northern Ireland, there is the Department of Agriculture. Each of the above can be contacted through local or regional offices. For private gardeners also, advice may be available from the above sources in special circumstances. In England and Wales, some help is given by Agricultural Institutes. Local authorities are becoming increasingly active in giving help and advice to private gardeners. So also are chemical firms, seed firms, the National Trust, and the larger horticultural societies – not forgetting gardening magazines and newspapers. Many more private consultants are now available throughout the country.

Bad Gardeners

A plant's very worst enemy can be the gardener, inquisitive or meddling, rough or careless, overlavish with water or fertilizer, or the complete converse, starving or depriving the needy plant, often until it dies a premature death. It is said of many gardeners that they have 'green fingers', their very touch coaxing the seed to germinate, the cutting to root and the plant to give of its best generally in terms of flower or fruit. Some other gardeners say of themselves that they are 'death to any living plant', causing it to wilt or perform badly. A little patience and sympathy for all living things, backed by sound common sense, is the explanation of 'green fingers'. Impatience, careless handling and poking around to see if cuttings have rooted is the very antithesis of green fingers. Most growing things resent constant disturbance, as checks to growth leave the plant in a state of shock, vulnerable to infection by weak parasites or diseases which it would shrug off contemptuously if growing normally. Infection can also result from damage caused by garden tools.

Apart from causing damage by over-watering, especially the young plant with minimal capacity for water uptake, gardeners are often guilty of killing plants by over-feeding. It is often not appreciated that a fertilizer can be as potent as a weedkiller if misused, especially in the vicinity of the soft stem of a young plant.

Treat plants as living things requiring a reasonable amount of food, water, light and warmth and the minimum of disturbance, and you will be well on the way to having a trouble free garden.

How to use this book

This book deals with the problems which commonly affect garden, greenhouse and commercial glasshouse plants. Plants, pests, diseases and other problems are listed alphabetically and within each plant entry are listed the pests and diseases which affect that particular plant. If you do not know the name of the trouble affecting your plant, look up the plant entry and look for the symptoms it shows to find the cause and the suggested control measures. If you suspect a particular pest or problem, look up the entry for that pest or problem. The information you find there will help you to establish what the problem is, and you will then be able to take the recommended control measures.

Problems

The problem entries include insect pests, fungal, bacterial and viral diseases as well as some other causes of damage. The entries for common insect problems, for example Aphis, describe the pest, outline the important stages in its life cycle, and describe the damage caused by the pest at each stage in its life cycle. Then follow the recommended control measures and any preventative measures which should be taken. The entries for diseases, for example the fungal disease Botrytis, describe the general symptoms to look for and then list the control measures which should be taken.

Plants

The plant entries begin by outlining the plant characteristics. Hardiness, evergreen or deciduous, annual, biennial or perennial, height, flowering time, flower colours and use are all indicated. Height is indicated by a code of which details are given below. The entry then lists and enumerates all the pests and problems which affect the plant, giving the symptoms and control measures, including which chemicals to use. In some instances, symptoms or control or both are cross-referred to another entry.

Height code

Trees   Large: over 18m/60ft. Medium: 10–18m/35–50ft. Small: 4.5–9m/15–30ft.
Shrubs   Large: over 3m/10ft. Medium: 1.5–3m/5–10ft. Small: 1–1.5m/3–5ft. Dwarf: 30–60cm/1–2ft. Creeping: under 10cm/4in.
Herbaceous plants   Tall: over 3m/10ft. Large: 1.5–3m/5–10ft.

Medium: 45cm–1.5m/18in–5ft. Small: 15–45cm/6–18in. Dwarf: 10–15cm/4–6in. Creeping: under 10cm/4in.

Chemicals
Chemical names for pesticides and fungicides etc. are given in the text, but to find out what to ask for in the shop, consult the Appendix to this book which lists the available proprietary brands of the chemical.

Where chemicals are referred to as commercially used or available, they are listed under the *Agricultural Chemicals Approval Scheme* in the booklets revised and produced each February by the United Kingdom Ministry of Agriculture, Fisheries and Food, and are available at divisional offices of the Ministry of Agriculture – or in Scotland at the Department of Agriculture for Scotland.

Cross-references
Plants, pests, diseases and other problems are cross-referred throughout. A cross-reference under a subject, e.g. under **Cissus, 1** Aphis (see Aphis) refers to the main entry for Aphis but where a cross-reference contains a number, e.g. **43** Powdery mildew (see 7 above Apple mildew) it refers to another section of the same entry.

Regulations regarding Pests and Diseases
There are many official regulations regarding the sale of plants affected by pests or diseases, and the destruction of plants which have contracted pests and diseases. For example, elms affected by Dutch elm disease must be destroyed throughout Britain. Certain pests and diseases are notifiable, that is, they must be reported to the Ministry of Agriculture. To find out the regulations in your area, it is necessary to contact the local officer of the Ministry of Agriculture, or the local horticultural adviser.

Acknowledgements
My eternal thanks to Pauline Dovaston without whose expert editorial help this book would not have seen the light of day.

My sincere thanks to Marjorie Clark and Garth Foster for specialist information.

A

Abies (see Silver firs)

Abutilon *Abutilon megapotamicum* and varieties and other *Abutilon* species. Greenhouse (or mild climate) evergreen and deciduous trees and shrubs. Height: trees small, shrubs medium. Flowering time: spring, summer or winter according to sort. Flower colours: orange, purple, red, yellow, white or parti-coloured. Foliage: variegated. Use: greenhouse borders trained to rafters, pillars or walls, greenhouse pot plants, summer bedding, outside in mild areas only.

Pests and Diseases. 1 Glasshouse whitefly *Trialeurodes vaporariorum* (see Whitefly). **2** Vine mealybug *Pseudococcus obscurus* (see Mealybugs). May attack established plants. **Symptoms:** large colonies of mealybugs on plants; twisted foliage. **Control:** spray with Oxydemeton-methyl or Malathion. **3** Virus, Abutilon mosaic virus (see Virus diseases). **Symptoms:** variegated leaf colour. **Control:** none used, as the variegation is decorative.

Acaricides Types of pesticide effective against mites, for example, glasshouse red spider mite. The only Acaricide available to amateur gardeners is Dicofol, usually in combined form. Surprisingly, some fungicides are quite effective against mites, for example, Dinocap for fruit tree red spider mite and Benomyl for glasshouse red spider mite, and a careful perusal of product labels may reveal a useful joint action. See Appendix p. 245–54 for list of chemicals.

Acer (see Maple)

Achimenes *Achimenes* species and varieties. Greenhouse herbaceous perennials, with swollen, underground tubers. Height: small. Flowering time: spring, summer. Flower colours: orange, purple, purple-blue, red, white. Use: greenhouse pot plant.

Pests and Diseases. 1 Chlorotic leaf spots (see Physiological disorders). May occur on established plants. **Symptoms:** pale, ring-like spots on the leaves. **Control:** avoid overhead watering, especially with cold water; water from below; shade from bright sunshine.

Aconitum   (see Monk's hood)

African Marigold   (see Marigold)

African Violet   *Saintpaulia ionantha* and varieties. Greenhouse herbaceous perennials. Height: dwarf. Flowering time: varies; may flower almost continuously, then have a resting period. Flower colours: cream, blue, lilac, pink, purple red or white. Foliage: variegated. Use: greenhouse and indoor pot plants.

Pests and Diseases. 1 Chlorotic leaf spots (see Physiological disorders). May attack established plants. **Symptoms:** irregular, yellowish rings and lines on leaves. **Control:** water from below; avoid overhead watering, particularly with cold water; shade plants from bright sun. **2** Crown rot, caused by the fungus *Pythium ultimum*. May attack established plants. **Symptoms:** wilting leaves and softening of plant. **Control:** destroy badly affected plants; do not overcrowd or overwater plants; take leaf cuttings from healthy plants only; use sterilized compost for rooting. **3** Cyclamen mite *Tarsonemus pallidus* (see Cyclamen 3 Cyclamen mite). **4** Fungus gnats *Bradysia* species. May attack at all stages. **Symptoms:** small black flies scuttling over pots and occasionally taking flight. The larvae are cylindrical, white with black heads. They burrow into the undersides of the leaves and damage the roots. **Control:** as a general precaution, keep sterilized soil in sealed bags, and avoid excessive moisture, which favours the growth of green, algal scum. *Saintpaulia* is a difficult subject to treat with pesticides, being easily scorched. A drench of Diazinon should be applied in dull conditions, tilting the plant to drain spray solution from the leaves. No pesticide should be applied when the plants are flowering. **5** Grey mould, caused by the fungus *Botrytis cinerea* (see Botrytis).

Aloe   *Aloe* species and varieties. Greenhouse, herbaceous, evergreen perennials. Height: dwarf to tall. Flowering time: summer. Flower colours: pink, red or yellow. Only a few species flower in Britain. Plants may have stems or may be stemless. Leaves: thickened, succulent, with sharply toothed margins and prickles, arranged in rosettes or spirals, very decorative, often variegated and patterned. Use: greenhouse borders, greenhouse pot plants.

Pests and Diseases. 1 Mealybug, Root mealybug *Rhizoecus elongatus* (see Mealybugs). May attack established plants. **Symptoms:** poor growth; small, yellow structures covered with white wax on the roots. **Control:** if possible, isolate affected plants; treat the soil with a dilute wash of Diazinon and repeat in 14 days.

Althaea   (see Hollyhock)

Alyssum   *Alyssum* species and varieties. Hardy and greenhouse annuals, perennials and sub-shrubs. Height: dwarf to small, sub-shrubs dwarf. Flowering time: spring, summer. Flower colours: pink, purple, white or yellow. Foliage: silvery grey. Use: annual borders, greenhouse pot plants, rock gardens, summer bedding, window boxes and outdoor containers.

Pests and Diseases. 1 Downy mildew *Peronospora galligena* (see Mildew). May attack established plants. **Symptoms:** curling of leaves with downy growth on the underside; acute distortion of the leaves. **Control:** spray with Thiram or Zineb, or dust with flowers of sulphur, at first sign of attack; use sterilized compost and clean containers.

Anemone   Du Caen or St Brigid anemone derived from *Anemone coronaria*, Japanese anemone *Anemone hupehensis japonica* and varieties and other *Anemone* species and varieties. Mostly hardy herbaceous perennials, with thickened rootstocks or tubers. Height: small to medium. Flowering time: spring, summer, autumn. Flower colours: blue, cream, pink purple, red, white or yellow. Use: borders, herbaceous borders, cut flowers, rock gardens.

Pests and Diseases. 1 Aphis, various species of *Aphis*. **Symptoms:** badly curled and distorted leaves. **Control:** (see Aphis). **2** Caterpillars (see Caterpillars). **3** Cluster cup, caused by the fungus *Puccinia pruni-spinosa* (see Rust). May attack established plants. **Symptoms:** orange-coloured pustules on undersides of the leaves. Badly affected plants seldom flower. **Control:** destroy affected plants immediately. **4** Downy mildew, caused by the fungus *Plasmopara pygmaea* (see Mildew). May attack established plants. **Symptoms:** white mould on the undersides of the leaves. A bad attack may be followed by mildew caused by the fungus *Botrytis cinerea*, which is similar in appearance. **Control:** spray with a copper fungicide, Thiram or Zineb. Benomyl is helpful in checking the onset of *Botrytis cinerea*. **5** Rust, caused by the fungus *Puccinia fusca* (see Rust). May attack established plants. **Symptoms:** chocolate-brown spots on the undersides of leaves; plants tend to grow taller than normal. This disease is unsightly rather than serious. **Control:** spray with Zineb. **6** Smut, caused by the fungus *Urocystis anemones*. May attack established plants. **Symptoms:** blisters and streaks on the leaves, which give rise to sooty black spores. Other plants, including wild buttercups, can act as hosts to this disease (see Host). **Control:** lift and destroy affected plants immediately.

Antirrhinum *Antirrhinum majus* and varieties. Hardy herbaceous perennials. Height: small to medium. Flowering time: summer outside; winter/spring when grown in the greenhouse. Flower colours: pink, red, white, yellow or parti-coloured. Use: summer bedding and borders, window boxes, outdoor containers, commercial glasshouse cut flowers and pot plants. Usually treated in the open as a half-hardy annual.

Pests and Diseases. 1 Aphis, Peach-potato aphis *Myzus persicae* (see Aphis). May attack at all stages. **Symptoms:** pink/yellow aphids on plants; stunted plants. **Control:** spray with Oxydemeton-methyl or Malathion. Commercially, Demeton-S-methyl may be used. **2** Cutworms (see Cutworms). **3** Downy mildew, caused by the fungus *Peronospora antirrhini* (see Mildew). May attack young and established plants. A serious disease. May be troublesome in the greenhouse. **Symptoms:** stunted plants; curled leaves, with grey felt-like growth on the undersides. **Control:** destroy initially infected plants; spray with a copper fungicide, Thiram or Zineb; avoid a cool, humid atmosphere in the greenhouse by ventilating and giving some warmth; use sterilized compost and containers. **4** Eelworm (see Eelworms). May attack established plants. **Symptoms:** puckered and distorted leaves. **Control:** remove and destroy plants; practise rotation (see Rotation). **5** Foot rot, caused by the fungus *Phytophthora cactorum* (see Foot rot). May attack young and established plants. **Symptoms:** wilting, with the fungus girdling the stems just above soil level, followed by death. This disease affects many different kinds of plants. **Control:** commercially use Milcol; use sterilized compost or soil-less media and clean containers (see Composts). **6** Leaf spot and Stem rot, caused by the fungus *Phyllosticta antirrhini* (see Leaf spot, Stem rot). May attack young and established plants. **Symptoms:** leaves show yellow spots, which may join together and kill the leaves; shoots attacked and may die back. Owing to improved methods of seed production, this disease, which is carried on the seed, is not common today. **Control:** do not save seed from infected plants; destroy diseased plants, as the fungus can overwinter on old plants; sterilize ground with Basamid. **7** Rosy rustic moth *Hydraecia micacea*. May attack young plants.

Symptoms: sudden wilting and death of plants. Medium-sized pink or red caterpillars may be found burrowing within the stem bases. The moth lays its eggs in the autumn and the young caterpillars hatch in the spring. **Control:** chemical control is usually too late to be effective because the damage happens without warning. Maintaining a weed-free plant bed in the autumn greatly reduces the risk of damage in the following spring, because the moths are not attracted to the beds. **8** Rust, caused by the fungus *Puccinia antirrhini* (see Rust). A serious disease. **Symptoms:** masses of chocolate-brown spores on the undersides of the leaves causing death of leaves and failure of plants to flower. In areas where the disease is severe, rust-resistant varieties only should be grown. **Control:** spray with Thiram or Zineb; dust with Oxycarboxin; practise rotation (see Rotation). **9** Sclerotinia rot, caused by the fungus *Sclerotinia sclerotiorum*. May attack established plants. **Symptoms:** rotting of shoots and flowers, which are covered with white fungal growth, later turning black. **Control:** spray with Bordeaux mixture; use sterilized compost and clean containers; practise rotation (see Rotation). **10** Shot hole, caused by the fungus *Heteropatella antirrhini*. **Symptoms:** pale green spots on the leaves drop out, leaving holes about 7 mm ($\frac{1}{4}$ in) across. **Control:** spray with a copper fungicide or, in serious cases, destroy the plants. **11** Slugs (see Slugs and snails). **12** Wilt, caused by the fungus *Myrothecium roridum* (see Wilt). **Symptoms:** white fungal growth at the bases of the stems, followed by wilting and, usually, death. This disease may attack other plants. **Control:** in the greenhouse, use sterilized compost, or soil-less media and clean containers (see Composts); outside, ground may be sterilized with Basamid (see Soil sterilization); try watering plants with Benomyl or commercially with Milcol; practise rotation (see Rotation).

Ants   The common ant *Lasius niger* and other species, which vary in colour and size, are very familiar, as are their large nests or heaps,

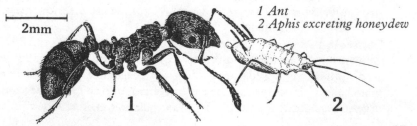

2mm

1 Ant
2 Aphis excreting honeydew

1   2

and the active processions of ants around nests. Ants themselves are not particularly injurious to plants, though they may disturb seeds and loosen roots. They may also spread pests, as they roam over plants, in their search for honeydew excreted by aphids. Ants may fly as well as creep, so may also enter greenhouses through ventilators.

Control: destroy nests by insertion of carbon disulphide or HCH smokes; kill ants with a proprietary ant killer containing for instance, Borax, Chlordane, Pyrethrum or Trichlorphon; use flame guns for local control; dig camphor into the soil; place grease bands on tree stems to stop ants climbing up.

Aphis Winged and wingless pests, commonly known as aphids or greenfly, which attack a wide range of plants at all stages of growth in the open and in the greenhouse. They may be green, black or some other colour. There are a number of different species, with life cycles which may be very complicated. Basically, they start life as eggs, laid on the host plant or on other plants (see Host). The eggs hatch into immature females, called nymphs, which grow in size, shedding their skins as they outgrow them, and finally becoming wingless adults. They are *viviparous*, producing living young, which in time produce more young. Finally, winged females are produced which can migrate to other plants and lay eggs there. Aphids feed by sucking the sap of plants, usually attacking young shoots and leaves, though some species may feed on older stems or on roots. Various parts of plants may be affected, and symptoms vary according to the species of aphids. Some aphids cause extreme leaf curling. Some cause distortion of shoots, stems and flowers. Some cause galls and swellings. Some infest roots. Aphids weaken plants by depriving them of valuable sap. They also open wounds which may allow certain diseases to attack. Aphids excrete honeydew which attracts sooty moulds and these weaken and disfigure plants (see Honeydew). Also, certain virus diseases are spread by aphids feeding on diseased plants and then carrying virus-infected sap to healthy plants. As aphids are a serious pest, they must be controlled.

Control: use one of the many different insecticides available, following manufacturer's directions; vary the kind of insecticide used, otherwise aphids may develop resistance to particular chemicals. It is often impossible to control aphids effectively on large trees, shrubs and herbaceous plants outside. Chemicals suitable for the amateur include Derris (Rotenone), Dimethoate, Formothion, Malathion, Oxydemeton-methyl, Pirimicarb and Pyrethrum. For commercial growers, the following chemicals are listed in the *Ministry of Agriculture Agricultural Chemicals Approval Scheme* booklet—Aldicarb, Demephion, Demeton-S-methyl, Dimethoate, Disulfoton, Formothion, Malathion, Menazon, Oxydemeton-methyl, Phorate, Phosphamidon and Thiometon.

Apple Apple varieties derived from *Malus pumilus* wild crab apple, and other *Malus* species. Hardy and greenhouse deciduous trees. Height: small to medium. Flowering time: spring. Flower colour: pink and white. Fruit colour: green, yellowish-green, may be flushed or streaked with red or yellow. Use: food crop grown for its fruit, may be grown in pots in the greenhouse. Where plants in the greenhouse may be attacked, this is noted.

Pests and Diseases. 1 Aphis (see Aphis). (a) Apple and grass aphid *Rhopalosiphum insertum*. (b) Green apple aphid *Aphis pomi*. (c) Plantain aphid *Dysaphis plantaginis*. (d) Rosy leaf-curling aphid *Dysaphis devecta*. (e) Woolly aphid (see 54 below Woolly aphid). May attack young and established plants. **Symptoms:** aphids and honeydew present on the plants (see Honeydew); distortion of leaves and shoots; red or yellow discolouration of leaves; leaf curling; restricted growth. **Control:** spray with winter wash, tar oil or DNOC; if bad attacks occur in spring or summer, spray with Derris, Dimethoate, Malathion or any other suitable insecticide; spray prior to flowering in spring. May also attack plants in the greenhouse. **2** Apple blossom weevil *Anthonomus pomorum*. May attack established plants. **Symptoms:** greyish adults, with V-shaped mark on their backs, pierce developing buds and lay eggs; legless whitish larvae feed inside the buds, causing them to turn brown and die before opening (buds in this condition are called capped buds); adult weevils leave buds, feed for a time on the leaves, biting pieces out of them, then move to winter shelter of tree bark, fence posts, hedge bottoms or rubbish heaps. **Control:** at bud breaking stage, spray with Carbaryl, Fenitrothion or gamma-HCH. **3** Apple and pear bryobia mite *Bryobia rubrioculus* (see Mites). May attack young and established plants. **Symptoms:** leaves and flowers marked by mites sucking sap from them. **Control:** spray with winter wash, tar oil or DNOC; in summer, spray with Derris, Dimethoate or a systemic insecticide (see Systemic chemicals).

4 Apple canker, caused by the fungus *Nectria galligena*. Attacks established plants. **Symptoms:** small sunken patches on branches or bark which enlarge into oval areas which crack open to form ragged wounds, causing eventual dieback; there may be whitish spores round the rims of the cankers. This is a serious disease, which enters through wounds of any kind. **Control:** avoid leaving large open pruning cuts and use canker paints after pruning; commercially, use DNC sprays plus mercury at the bud break stage, followed by 2 mercury sprays in the dormant period. May also attack plants in the greenhouse. **5** Apple capsid *Plesiocoris rugicollis*. May attack young and established plants. **Symptoms:** green adults on trees June/July feeding on the leaves, making small holes, and laying eggs; the following spring, small yellow capsids hatch out, run over the leaves and feed by sucking sap, making brown spots on the leaves in April/May; later they feed on the developing fruitlets, causing scarring which develops into large corky scars. **Control:** spray at the green and pink bud stage with Dimethoate or Fenitrothion; repeat if necessary at petal fall. May also attack plants in the greenhouse. **6** Apple leaf midge *Dasineura mali*. May attack young and established plants. **Symptoms:** a certain amount of leaf rolling. **Control:** spray with Carbaryl in May, and as there are several generations, repeat between the end of June and the beginning of July and again during August. **7** Apple mildew, caused by the fungus *Podosphaera leucotricha*. May attack established plants. **Symptoms:** leaves, blossom trusses, and new shoots affected by fluffy white fungal mould, which causes distortion and checks growth. The disease overwinters in the fruit buds, ready to start new attacks in the following spring. Complete control is difficult. **Control:** prune out badly mildewed shoots in March; spray with Dinocap or a systemic fungicide (see Systemic chemicals) such as Thiophanate-methyl at the pink bud stage (late April to early May); follow up with further regular sprays at intervals of around 7–14 days. May also attack plants in the greenhouse. **8** Apple sawfly *Hoplocampa testudinea*. May attack established plants. **Symptoms:** small yellowish-red insects with 4 wings, lay eggs just below the flower calyx and the small whitish caterpillars which hatch out bore into the side of the fruit, often moving along beneath the skin and causing ribbon-like scars; they feed on and around the core, leaving a black messy material called 'frass'; affected fruit smells unpleasant and usually drops in early summer. **Control:** a wide range of chemicals may be used, including HCH and systemic insecticides (see Systemic chemicals) about 7 days after petal fall. (For the difference between apple sawfly and codling moth damage, see 25 below Codling moth.) **9** Apple scab, caused by the fungus *Venturia inaequalis*. May attack young and established plants. **Symptoms:** sooty blotches on the leaves; blistering on twigs; dark scabs and spots and cracking on fruit; fruit in store may have black, sunken, saucer-shaped spots. This is a very common trouble especially in wet seasons, wet areas and if plants are overcrowded and congested. **Control:** collect and burn fallen leaves; prune and burn diseased twigs before bud-burst; spray with Benomyl, Bordeaux mixture, Captan, lime sulphur, Thiophanate-methyl or other recommended fungicide at regular intervals from bud-burst onwards; check that varieties are not liable to damage before using copper or sulphur sprays; note that some varieties are more susceptible to scab than others so avoid these in areas where scab tends to be bad. May also attack plants in the greenhouse. **10** Apple sucker *Psila mali*. May attack established plants. **Symptoms:** winged adult insects, not unlike aphids, may be seen during the summer on neglected plants; minute, elongated, straw-coloured eggs are laid on fruit spurs; in spring, yellowish-white nymphs (immature form) hatch out and suck sap from the developing buds, causing malformation of leaves and fruit; a waxy sticky excretion may be seen on infected fruit clusters. **Control:** winter wash adequately controls this pest but, if not used, spray with gamma-HCH, Malathion or Nicotine. **11** Birds (see Birds). **12** Birds, Blackbirds, Thrushes, Woodpigeons (see Birds). **13** Bitter pit (see Physiological disorders). May occur in fruit. **Symptoms:** brown, slightly sunken speckles on the skin, with brown areas scattered through the flesh, making the fruit bitter and useless. Cause not known exactly, but may be due to calcium deficiency (see Physiological disorders). **Control:** commercial growers may spray with calcium nitrate at 10 kg per hectare (10 lb per acre), possibly giving several applications from late June until September; try to avoid trees suffering from drought in summer by watering and mulching, if practical; if trees have suffered badly from bitter pit, avoid heavy pruning, which can cause this trouble. **14** Bitter rot (see 37 below Gloeosporium rots, Bitter rots, Storage rots). **15** Blossom wilt (see 16 below Brown rot and blossom wilt). **16** Brown rot and blossom wilt, caused by the fungi *Sclerotinia fructigena, Sclerotinia* *laxa f. mali*. May attack established plants. **Symptoms:** withered flower trusses and fruit spurs (blossom wilt stage); fruits bearing greyish or buff-coloured cushions of fungal spores, often in concentric circles; fruits later mummified (brown rot stage); cankers on branches. **Control:** cut out cankers and infected spurs during spring or early summer when infection is visible; remove and destroy infected and mummified fruits, both on and off trees; pre-blossom spraying with Benomyl or lime sulphur (except on sulphur-shy varieties), together with tar oil winter washes, will reduce blossom wilt; control insects, which may produce entry points for the disease; spray with Orthocide (Captan) as a preventative measure; use Thiophanate-methyl as a routine measure to reduce the risk of infection. May also attack plants in the greenhouse. **17** Bryobia mite (see 3 above Apple and pear bryobia mite). **18** Bud moth *Spilonota ocellana*. May attack established plants. **Symptoms:** caterpillars bore into fruit buds and later feed on leaves and flower trusses. **Control:** (see 35 below Fruit tree tortrix moths). **19** Canker (see 4 above Apple canker). **20** Caterpillars (see 8 above Apple sawfly, 18 above Bud moth, 22 below Cherry bark tortrix moth, 24 below Clouded drab moth, 25 below Codling moth, 28 below Dock sawfly, 32 below Fruitlet mining tortrix moth, 35 below Fruit tree tortrix moth, 51 below Summer fruit tortrix moth, 53 below Winter moth). **21** Chafer beetles, Cock chafer *Melolontha* species, Garden chafer *Phyllopertha* species. May attack established plants. Cock chafers are brown beetles about 25 mm (1 in) long. Garden chafers are smaller and metallic blue-green in colour. **Symptoms:** beetles bite pieces out of the fruitlets in early summer, completely spoiling them; white larvae, which live in the soil, feed on the roots. **Control:** spray with Carbaryl in May or June as soon as the beetles are seen. **22** Cherry bark tortrix moth *Enarmonia formosana*. May attack established plants. The caterpillars are dull pinkish-white or dirty brown with black heads, and are about 12 mm ($\frac{1}{2}$ in) long. **Symptoms:** caterpillars bore into trunks near ground level and form tunnels; caterpillars feed on the leaves and on the fruit. **Control:** scrape away the bark and paint with tar oil wash in March. **23** Clay-coloured weevil *Otiorhynchus singularis* (see Weevils). May attack newly grafted, young, or established plants. **Symptoms:** grafted areas are damaged by weevils gnawing; weevils feed by night, so are not usually seen; shoots of established plants may also be attacked. **Control:** spray with Carbaryl at the base of the trees. **24** Clouded drab moth *Orthosia incerta*. Moths are deep pinkish-brown and caterpillars are light green with a long greenish-yellow line along their backs. **Control:** spray at petal fall with Chlorpyrifos commercially. **25** Codling moth *Cydia pomonella*. May attack fruit. The adult moths are pinkish brown, fly at dusk, and are noticeable on warm sultry evenings. When fully grown, the caterpillars are pinkish with brown heads and scattered hairs, and are 12 mm ($\frac{1}{2}$ in) long. **Symptoms:** eggs are laid on leaves, shoots, spurs and fruitlets; young caterpillars enter fruit by the side or by the eye (stalk end); they burrow and feed round the core, eating pips and flesh; they leave the fruit by the side; fruit may be still attached or may have fallen; caterpillars spend the winter as cocoons on the bark or in any dry rubbish; moths emerge in spring. **Control:** start spraying in mid-June and repeat twice at 3-weekly intervals if necessary, using Carbaryl, Fenitrothion or Malathion. Note that the damage done by Apple sawfly (see 8 above) and Codling moth is often confused. Apple sawfly caterpillars are creamy white, codling moth caterpillars are larger and are pale pink. Damage by sawfly caterpillars is done earlier in the season. Fruit attacked by sawfly smells unpleasant, but no smell with codling attacks. Apple sawfly caterpillars make ribbon-like scars on the fruit, codling caterpillars do not. Sawfly caterpillars eat large areas inside the apple while codling caterpillars feed on the cores. **26** Collar rot, caused by the fungi *Phytophthora cactorum, Phytophthora syringae*. May attack established plants. **Symptoms:** foliage turns yellow; tissue at the base of tree stained orange; fruit remains small. **Control:** clear away soil and cut away infected tissue; paint wounds with a canker paint; keep bases of trees weed-free. **27** Cox spot (see Physiological disorders). May occur on established plants. **Symptoms:** dead brown patches on the leaves of Cox's Orange Pippin and some other varieties. This trouble is mainly due to heavy rainfall. **Control:** no control. New foliage will quickly develop. **28** Dock sawfly *Ametastegia glabrata*. May attack established plants. Caterpillars are apple green in colour, and 12 mm ($\frac{1}{2}$ in) long. **Symptoms:** caterpillars first feed on weeds and later bore into fruit. **Control:** spray with Carbaryl; control weeds. **29** Earwig *Forficula auricularia*. May attack fruit. **Symptoms:** earwigs puncture fruit and hollow out centres. **Control:** spray with Carbaryl or gamma-HCH. **30** Fire blight

Erwinia amylovora (see Bacterial diseases). May attack established plants. **Symptoms:** growing points show reddish colouration. **Control:** none necessary. This disease is not usually of great concern. **31 Frost damage.** May be confused with damage caused by various pests and diseases. Damage caused by frost shows as the leaves develop. **Symptoms:** the undersides may be blistered and cracked, the leaves round flower trusses usually being the worst affected; the leaves do not drop and new leaves develop; very severe frost may destroy the blossom. **32** Fruitlet mining tortrix moth *Pammene rhediella*. May attack established plants. **Symptoms:** yellow caterpillars feed on leaves or bore into developing fruit; damage on fruit is usually followed by brown rot (see 16 above Brown rot and blossom wilt). **Control:** spray with Fenitrothion in early June. **33** Fruit tree red spider mite *Panonychus ulmi* (see Mites). May attack young and established plants. Mites may be seen with a hand lens. **Symptoms:** mites feed on the leaves, giving them a silvery appearance, and greatly reducing their efficiency. Control is made difficult by the fact that strains of this mite develop resistance to various chemicals (see Strain). **Control:** spray with DNOC at bud break; after petal fall various sprays can be used, including Chlorpyrifos, Dichlorvos or Dinobuton commercially, but amateur gardeners will more usually spray with Derris, Dicofol (combined form), Dimethoate, Malathion or other acaricides on a varied basis. May also attack plants in the greenhouse. **34** Fruit tree rhynchites, Twig cutting weevil *Rhynchites coeruleus* (see Weevils). May attack established plants. **Symptoms:** blue weevils feed on foliage in spring; eggs are laid on young shoots and weevils then cut the shoot just below the eggs; shoots may hang for a while, but eventually fall; larvae feed on pith of shoots, overwinter in the soil and adults emerge in spring. This pest is not widespread. **Control:** spray with Carbaryl at the pink bud stage. **35** Fruit tree tortrix moth *Archips podana, Ditula angustiorana* and others. May attack established plants. Caterpillars are pale yellow and, early in the season, quite small. **Symptoms:** caterpillars feed first on the leaves and then on the fruitlets, causing surface marking. **Control:** use Fenitrothion at the green bud stage, coupled with normal spraying as for Codling moth in summer, to control the caterpillars generally; apply winter washes, which are usually more effective if they contain a general insecticide. May attack plants in the greenhouse. **36** Glasshouse red spider mite *Tetranychus urticae*.

May attack young and established plants in the greenhouse. **Symptoms:** (see Tomato 18 Glasshouse red spider mite). **Control:** spray with Dimethoate or Demeton-S-methyl, Tetradifon commercially in mid-May, and again 3 weeks later. **37** Gloeosporium rots, Bitter rots, Storage rots, caused by the fungi *Gloeosporium album, Gloeosporium perennans*. May attack fruit in store, more usually fruit from older trees. Infection may start before harvesting. **Symptoms:** mushy brownish saucer-shaped areas on the fruit, with white spores forming. **Control:** inspect and then store fruit correctly and inspect it regularly in store; remove and destroy any fruit showing signs of disease; spray in August with Thiophanate-methyl and repeat shortly before harvesting; Captan is also effective to a certain extent; avoid excessive use of nitrogen fertilizers; avoid pruning too early in spring. **38** Leaf and bud mite *Phyllocoptes schlechtendali*. May attack established trees. Mites are brown in colour and difficult to see. **Symptoms:** damage similar to Fruit tree red spider mite damage, with silvering of the foliage, coupled with blisters on the upper surfaces of the leaves. **Control:** after petal fall, use Dicofol as required (see 33 above Fruit tree red spider mite). **39** Leaf hoppers. May attack young and established bushes. **Symptoms:** bleached mottling of the leaves caused by small, quick-moving greenish insects feeding on the undersides; cast white skins on the leaves. Not a serious pest in itself but may spread virus diseases. **Control:** spray with Carbaryl in June and repeat 3 weeks later; Dimethoate and Malathion may also be used; use tar oil winter wash. **40** Mildew (see 7 above Apple mildew). **41** Nectria eye rot, caused by the fungus *Nectria galligena*. May attack fruit in store. **Symptoms:** rotting similar to that of Gloeosporium rot, but usually starting round the stalk ends of the fruit. **Control:** (see 37 above Gloeosporium rots, Bitter rots, Storage rots). **42** Physiological disorders (see Physiological disorders). **Symptoms:** various symptoms, such as brown spotting of the flesh, and leaf discolouration, which may be due to mineral deficiencies. Usually due to unsuitable soil, lack of feeding or bad drainage. **Control:** improve growing conditions; in certain cases, foliar feeds may be used. **43** Powdery mildew (see 7 above Apple mildew). **44** Pre-harvest drop (see Physiological disorders). **Symptoms:** fruit falls before it is fully ripe, especially during windy weather. **Control:** spray with ANA commercially before the period when fruit drop usually occurs, which varies according to variety. **45** Rabbit (see

Rabbit). **46** Red spider (see 33 above Fruit tree red spider mite, 36 above Glasshouse red spider mite). **47** Sawfly (see 8 above Apple sawfly, 28 above Dock sawfly). **48** Scab (see 9 above Apple scab). **49** Silver leaf, caused by the fungus *Stereum purpureum*. May attack established plants. **Symptoms:** silvering of the leaves. **Control** (see Plum 15 Silver leaf). **50** Storage rots (see 37 above Gloeosporium rots, Bitter rots, Storage rots). **51** Summer fruit tortrix *Adoxophyes orana*. **Symptoms** and **Control** (see 35 above Fruit tree tortrix moth). But damage occurs later and caterpillars go deeper into the fruit after feeding on the leaves. **52** Weevils (see 2 above Apple blossom weevil, 23 above Clay-coloured weevil). **53** Winter moth *Operophtera brumata*. May attack young and established plants. The mature caterpillars are 37 mm (1½ in) long, green in colour with pale lines along the body. **Symptoms:** looper type caterpillars hatch in spring and may do serious damage in feeding on young leaves and blossom and may later attack young fruitlets. **Control:** spray at green and pink bud stage using Chlorpyrifos or Fenitrothion; repeat after blossom if necessary. **54** Woolly aphid *Eriosoma lanigerum*. May attack young and established plants. **Symptoms:** stems and branches may be crowded with aphids which secrete a waxy white material called wool. Not a serious pest in themselves, but they cause corky scars and galls which allow entry of disease spores, such as apple canker spores. **Control:** Chlorpyrifos is recommended commercially, but if on a small scale, clear off aphids and wool with a brush and methylated spirit.

Apricot Varieties derived from *Prunus armeniaca*. Greenhouse (outdoors in milder areas only) deciduous trees. Flowering time: spring. Flower colours: pink or white. Fruit: yellow, flushed red. Use: food crop, grown for the fruit, trained against walls in the greenhouse, against walls and sometimes as trees in the open. For **Pests and Diseases**, see Peach.

Approved stock Many groups of plants, largely but not exclusively, those which have been propagated vegetatively by cuttings, layers, grafting, budding or means other than by seed, can, over the years, lose vigour. This is due not only to genetical causes, but also to a build-up of pests and diseases, especially of virus diseases (see Virus diseases). The approval of stock (see Stock) by various research groups and bodies, for example the National Seed Development Organisation and the Nuclear Stock Association ensure that only disease-free selected stocks of a range of plants are distributed, whether or not the plants in question are subject to a certification scheme (see Certified stock). A wide range of plants has been rendered disease-free by special propagation methods, heat treatments (see Heat treatment) and other means.

Aquilegia (see Columbine)

Armillaria mellea This fungus is known by various names, including bootlace fungus, collar rot fungus, honey fungus, root rot fungus, shoestring fungus and tree root rot fungus. It is one of the most common and most destructive of tree fungal diseases. It can affect trees and shrubs of all kinds, and may even attack herbaceous plants. The fungus lives on dead trees, tree stumps and roots, and on other rotted material. From such rotted material it can travel rapidly underground, by means of bootlace-like strands called rhizomorphs, to attack the roots of living trees, being especially damaging to the roots of young trees. The roots of affected trees start to rot, and masses of white, fungal strands may be seen if part of the tree bark, which will come away easily, is raised at soil level. The honey-coloured, bracket-like toadstools seen growing out of dead trees and tree stumps, are usually the fruiting bodies of *Armillaria mellea*, though other fungal diseases which attack trees may produce similar growths.

1 Fruiting bodies   2 Rhizomorphs

Control: obviously clearing ground is not enough; it is essential to remove and burn all infected material before planting young trees; check drainage and soil, as poor growing conditions will make trees more liable to attack; to stop the spread of the rhizomorphs from an infected tree or from an infected area of ground, dig a trench 45 cm (18 in) wide, and 60–90 cm (2–3 ft) deep round it, throwing the soil inwards, so as to prevent any infected material from falling outwards on to clean ground. Many authorities claim that there is no really effective control other than the removal of affected trees, and, if possible, the treatment of the ground with formaldehyde or Basamid (see Soil sterilization). Phenolic emulsion, containing a specific percentage of phenol, has recently been developed for the treatment of affected trees.

Arum lily *Zantedeschia aethiopica* and varieties and other *Zantedeschia* species. Hardy in mild areas but best planted in water, and greenhouse herbaceous perennials with thick rhizomes or corms. Height: small to medium. Flowering time: varies from spring, summer to early autumn. Flower colours: white, greenish-white, pink, violet-red or yellow; the showy, coloured part of the flower is a bract, or modified leaf. Foliage: decorative, glossy and sometimes mottled. Use: greenhouse pot plants, commercial cut flowers in the glasshouse, water gardens.

Pests and Diseases. 1 Corm rot *Bacterium aroideae, Erwinia carotovora* (see Bacterial diseases). May attack established plants. **Symptoms:** distorted leaves; whitish spots and streaks on the foliage, which later turn brown; misshapen flowers. **Control:** in dormant season, wash corms; cut out and destroy diseased parts; soak in 2% formalin solution for 30 minutes; control thrips, which spread the disease, with insecticides. **2** Leaf spot, caused by various fungi (see Leaf spot). May attack established plants. **Symptoms:** wilting and death of plant; soft rot of corm. **Control:** destroy infected plants; thoroughly clean and disinfect benches on which plants will stand; wash down or fumigate greenhouse. **3** Spotted wilt, caused by Tomato spotted wilt virus (see Virus diseases). A serious disease. May attack young and established plants. **Symptoms:** white markings on leaves and stems; distortion of leaves; rotting of flower buds and flowers. **Control:** remove diseased leaves and flowers regularly; reduce humidity; control insects which spread the disease, especially aphis (see Aphis).

Ash *Fraxinus excelsior* and other *Fraxinus* species. Hardy deciduous trees and shrubs. Height: small to large. Flowering time: early summer. Flower colours: white, or flowers insignificant. Fruit: decorative and winged. Use: planted singly or in groups. Need ample space.

Pests and Diseases. 1 Bacterial canker *Pseudomonas savastanoi* (see Bacterial diseases). **Symptoms:** large cankers on branches and main trunk, deeper and more extensive than those of *Nectria galligena* (see 2 below Canker). **Control:** cut out and burn affected parts; seal with tar. **2** Canker, caused by the fungi *Nectria galligena,*

Phomopsis controversa, Phomopsis scobina. May attack established plants. **Symptoms:** large cankers on branches and main trunk; red resting bodies of the fungus on the diseased areas; death of buds. **Control:** cut or clean out affected area and paint with Arbrex (bitumen based), commercially Arbrex 805 (mercury based). The disease is more troublesome in wet conditions. **3** Dieback, caused by the fungus *Botrytis cinerea* and other species of fungi (see Botrytis). **Symptoms:** dieback of branches, especially when water collects at a crutch. **Control:** clean out and seal with cement if trouble has not progressed to an advanced stage. **4** Internal rots, caused by a number of different fungi, including honey fungus (see Armillaria mellea). **Symptoms:** dieback of branches; internal rotting of trunk or main branches; yellow or red-brown bracket-shaped fungal growths. **Control:** cut out affected branches or clear out affected areas as far as possible and seal wounds with cement, making it flush with the surface, so that the bark can later seal over the wounds.

Asparagus Varieties derived from *Asparagus officinalis*. Hardy herbaceous perennial. Height: medium. Ferny foliage consists of modified branches. Use: food crop, grown for its young shoots, cut in spring.

Pests and Diseases. 1 Asparagus beetle *Crioceris asparagi*. May attack established plants. **Symptoms:** adult beetles about 7 mm ($\frac{1}{4}$ in) long seen on foliage. Beetles have black heads, red bodies, yellow spots on black wing cases. Eggs laid on shoots and later foliage from June onwards. Usually two generations a year. Both beetles and larvae feed on shoots and foliage and may be damaging. **Control:** hand pick if on a small scale, otherwise spray regularly with Carbaryl, Derris or gamma-HCH from June onwards. Clear up rubbish from beds in winter. **2** Asparagus rust, caused by the fungus *Puccinia asparagi* (see Rust). May attack established plants. **Symptoms:** reddish pustules on all growth, seriously affecting shoots; black resting spores of the fungus may be seen in autumn on dead shoots. **Control:** spray with Maneb, Thiram or Zineb at the first sign of disease; cut down and burn foliage before resting spores form. **3** Violet root rot, caused by the fungus *Helicobasidium purpureum*. May attack established plants. **Symptoms:** poor growth; purple coloured roots. This disease may attack various root crops. **Control:** get rid of the plants and start a new bed on good ground that has not recently been used for root vegetables.

Asparagus fern *Asparagus plumosus, Asparagus sprengeri* and varieties. Greenhouse herbaceous perennials. Height: small or climbing. Foliage: decorative, clear green and feathery. Use: greenhouse pot plants and border plants, indoor pot plants, commercial glasshouse cut foliage.

Pests and Diseases. 1 Basal rot, caused by a species of *Pythium* fungus. May attack established plants. **Symptoms:** rotting of plants at ground level. **Control** (see Damping off, Root rot, Stem rot). **2** Root rot, caused by a species of Fusarium fungus (see Root rot). May attack established plants. **Symptoms:** drying and yellowing of plants; withering and death of basal shoots. **Control:** discard affected plants; use sterilized compost and clean containers; if plants are grown in the greenhouse border, sterilize soil, if possible by steaming. **3** Yellowing and leaf dropping (see Physiological disorders). May attack established plants. **Symptoms:** foliage turns yellow and drops. May be due to under-watering in summer, over-watering in winter, too dry an atmosphere, or too much lime in the compost or border soil. **Control:** correct general growing conditions.

Aspidistra *Aspidistra lurida.* Greenhouse, but will stand tempera-ture down to 5°C (40°F), herbaceous perennial. Height: medium. Leaves: long, stiff, dark green, and pointed. Use: greenhouse and indoor pot plant.

Pests and Diseases. 1 Leaf spot, caused by various species of fungi (see Physiological disorders). **Symptoms:** brown spots and marking on leaves. May attack established plants. May be due to over-watering or to too strong sunlight. **Control:** correct growing conditions; if severe, spray with copper fungicide.

Aster (see also Michaelmas daisy)   *Callistephus chinensis* China aster varieties. Hardy annual. Height: small to medium. Flowering time: summer. Flower colours: blue, pink, purple, red or white. Use: annual borders, summer bedding, commercial and garden cut flower in the open, commercial glasshouse cut flower, greenhouse and commercial glasshouse pot plant.

Pests and Diseases. 1 Aphis, Peach-potato aphis *Myzus persicae* (see Aphis). May attack young and established plants. **Symptoms:** distorted and withered leaves; aphids on the plants. **Control:** spray with Oxydemeton-methyl or Malathion. **2** Damping off, root and stem rots caused by the *Phytophthora* species of fungi, *Rhizoctonia solani* and *Thielaviopsis basicola* (see Damping off, Root rot, Stem rot). May attack at seedling stage especially, and also young plants.

Symptoms: collapse at soil level and death. **Control:** water the soil with Cheshunt compound, a copper fungicide or Zineb; use sterilized compost and clean containers. **3** Virus (see Virus diseases). May attack young and established plants. **Symptoms:** mottled, stunted and distorted growth; shoots die back. **Control:** spray with appropriate insecticide to destroy insects which spread the virus. Some varieties show virus resistance and plant breeders are aiming at immune varieties. **4** Wilt, caused by the fungi *Fusarium oxysporum* fusarium species *callistephi* and *Verticillium albo-atrum*. May attack at all stages. **Symptoms:** blackening of stem bases with *Fusarium*, similar but less discolouration with *Verticillium*; wilting of plants from the base upwards and yellowing of leaves. **Control:** use sterilized compost or soil-less media if disease is suspected (see Composts); practise rotation in the open (see Rotation). Wilt-resistant varieties are now available.

Aubergine Varieties of *Solanum melongena ovigerum*. Greenhouse annual. Height: medium to large. Flowering time: summer. Flower colour: blue. Fruit: cream or purple, with smooth shining skin, decorative. Use: food crop grown for the fruit, may also be grown for decoration.

Pests and Diseases. Aubergines are subject to many of the troubles which affect tomatoes (see Tomato). **1** Glasshouse red spider mite. May be troublesome (see Tomato 18 Glasshouse red spider mite).

Azalea (see Rhododendron)

B

Bacterial diseases Although common in man and animals, these are uncommon in plants. Bacteria are microscopic organisms, lacking chlorophyll, the green substance which enables plants to make food supplies, and therefore are dependent for food on living or dead organic matter. Some groups of bacteria are essential to life, some are useful and some cause diseases in man, animals and plants.

Symptoms are often soft spotting of leaves, or soft, slimy rotting of stems or bulbs for example, along with obviously unhealthy growth. Poor or abnormal growth or death of plants however, may be due to a number of different factors, such as attack by pests, or infection by bacterial, fungal or virus diseases, or by physiological disorders (see Fungal diseases, Virus diseases, Physiological disorders).

Control: as bacteria are present inside the plant tissues, sprays are not generally effective; remove and destroy all diseased plants; thoroughly clean benches and containers; use sterilized soil or composts; take care to propagate and plant only healthy plants; do not save seed from infected plants as in some cases seed carries bacteria.

Bean, Broad Varieties derived from *Vicia faba*. Hardy annual. Height: small to medium. Flowering time: spring to summer. Flowers: grey-white, marked with blue-black; fragrant. Large, thick pods. Use: food crop, grown for its seed.

Pests and Diseases. 1 Aphis, Blackfly, Black bean aphid *Aphis fabae* (see Aphis). **Symptoms:** black aphids clustering round growing points in early summer. Very prevalent, and damaging as growth is restricted. **Control:** at first signs of attack, spray with Derris or Malathion and repeat as necessary; in gardens, pinch out growing tips when plants have about four clusters of flowers; commercially, a number of different insecticides may be used, including Dichlorvos, Dimethoate, Formothion and Menazon. **2** Bean beetles, Pea and bean beetles *Bruchus rufimanus* and other *Bruchus* species. **Symptoms:** holes in seeds, made by larvae eating into the seeds in pods. Holes may be noticed when buying bean seed. **Control:** spray with gamma-HCH or Malathion at the flowering stage. **3** Bean weevils, Pea and bean weevils *Sitonia lineata* and other *Sitonia* species. May attack at all stages. **Symptoms:** scalloping of leaf margins where weevils have been feeding. Except for a bad attack at the seedling stage, not usually serious. **Control:** spray with Fenitrothion, gamma-HCH or Malathion. **4** Chocolate spot, caused by the fungus *Botrytis fabae* (see Botrytis). May attack established plants. **Symptoms:** brown blotches on leaves later spreading to pods. This disease does not cause much damage to the bean seeds. **Control:** spray with Benomyl, Bordeaux mixture or Thiram. **5** Damping off and foot rot, caused by *Fusarium* species of fungi (see Damping off, Foot

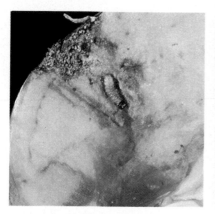

Apple **Codling moth** *Cydia pomonella* eating apple flesh (see page 23)

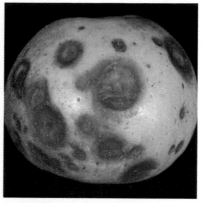

Apple **Gloeosporium rot** caused by *Gloeosporium album* and *Gloeosporium perennans*, mushy brown saucer-shaped areas on the fruit (see page 25)

Apple **Rosy leaf-curling aphid** *Dysaphis devecta*, distortion and red discolouration of leaves (see page 19)

Apple **sawfly** *Hoplocampa testudinea*, caterpillar feeding around the core, leaving behind a black messy material called frass (see page 20)

Apple scab caused by the fungus *Venturia inaequalis*, early stages showing dark scabs and spots and cracking on fruit (see page 21)

Apple **Woolly aphid** *Eriosoma lanigerum*, waxy white material called wool secreted by aphids (see page 26)

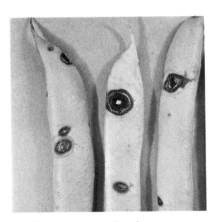

Bean runner or Scarlet runner **Anthracnose** caused by the fungus *Colletotrichum lindemuthianum*, dark spots on pods (see page 34)

Blackcurrant **Blackcurrant gall mite** *Cecidophyopsis ribis*, swollen and rounded buds caused by mite infestation (see page 44)

rot). **Symptoms:** plants rotting off at soil level, or bean seeds rotting in the ground *before* germination, especially in wet soil. **Control:** spray with Thiram; when replanting, drain ground well and see that it is in good condition; practise rotation and after a bad attack, sterilize soil chemically (see Rotation, Soil sterilization); use seed dressings Captan or Thiram (see Seed Dressing). **6** Leaf spots, caused by various fungi including *Ascochyta* and *Cercospora* species. May attack established plants. **Symptoms:** spotting and marking on leaves. **Control:** unnecessary, seldom causes serious damage. **7** Rust, caused by the fungus *Uromyces fabae* (see Rust). May attack established plants. **Symptoms:** orange pustules on leaves. **Control:** unnecessary, seldom causes serious damage.

Bean, Dwarf, French or Kidney Varieties derived from *Phaseolus vulgaris*. Half-hardy and greenhouse annual. Height: small. Climbing. Flowering time: summer. Flower colours: lilac or white. Long, flat, slender pods. Use: food crop, grown for its pods, outdoors and in greenhouse. For outdoor Pests and Diseases, see Bean, runner.

Pests and Diseases (greenhouse). **1** Aphis (see Aphis). May attack established plants. **Symptoms:** foliage curled or distorted; flowers misshapen. **Control:** use gamma-HCH smoke, gamma-HCH, Dimethoate or Malathion sprays. **2** Foot rot, caused by the fungus *Fusarium solani* fusarium species *phaseoli* (see Foot rot). May attack young and established plants. **Symptoms:** stem bases shrivelled and brown; leaves show yellowing which moves up the plant by degrees. **Control:** remove and destroy affected plants; do not grow beans in the same soil for several years, or if practical, sterilize the soil thoroughly before replanting in the following year. **3** French fly *Tryophagus neiswanderi* (see Cucumber 5 French fly). **4** Glasshouse red spider mite *Tetranychus urticae* (see Cucumber 9 Glasshouse red spider mite). **5** Glasshouse symphilid *Scutigerella immaculata* (see Tomato 19 Glasshouse symphilid). **6** Halo blight *Pseudomonas phaseolicola* (see Bacterial diseases). May attack established plants. **Symptoms:** water-soaked spots with yellow haloes on the leaves; spots may coalesce to form dead, brown areas; pods may be similarly affected, and seeds inside may be blistered. **Control:** remove and destroy the affected plants; avoid overhead watering, which spreads the bacteria; spray with Bordeaux mixture; do not sow blistered seed; resistant varieties are available (see Resistant varieties). **7** Red spider mite (see 4 above Glasshouse red spider mite). **8** Silver Y-moth *Plusia gamma*. **Symptoms:** caterpillars with green bodies and dark brown heads feed on plants during July, eating holes in the leaves, and leaving excrement granules, which are noticeable on the lower leaves. **Control:** spray with Fenitrothion or gamma-HCH. **9** *Symphilid* (see 5 above Glasshouse symphilid).

Bean, Runner or Scarlet runner Varieties derived from *Phaseolus coccineus*. Half-hardy herbaceous perennial, with thickened roots. Annual. Height: climbing. Flowering time: summer. Flower colours: red, white or parti-coloured. Long, flat pods. Use: food crop, grown for its pods.

Pests and Diseases. 1 Anthracnose, caused by the fungus *Colletotrichum lindemuthianum*. May attack established plants. A serious disease. **Symptoms:** dark spots on stems and leaves, becoming sunken and edged with red on stems; dark spots on pods, becoming sunken, centres pink and margins red. In severe cases pods may be covered with spots, the fungus may penetrate pods and infect seeds. **Control:** no really effective control. Always buy seed from a reliable source, as infected seed will spread the disease. **2** Aphis (see Bean, Broad 1 Aphis). **3** Bacterial blight *Xanthomonas phaseoli* (see Bacterial diseases). May attack established plants. **Symptoms:** small irregular amber-coloured areas, at first having a water-soaked or soft appearance, but gradually covering whole plants; the seed is infected and blotched with yellow or entirely yellow. **Control:** none possible, but obtain seed free of the disease. **4** Bean seed fly *Delia platura*. May attack early sown seed in cool springs, and young plants. **Symptoms:** larvae tunnelling into germinating seeds and later into stems. This increasingly prevalent pest is found mainly where plant refuse has not been removed. **Control:** ensure compost is well rotted and dug in; pre-soak seeds to promote germination; use cloches to force plants through the period when they are liable to attack. A top dressing of ammonium nitrate after germination will have the same effect. **5** Botrytis, caused by the fungus *Botrytis cinerea*. May attack established plants. **Symptoms:** blackening of leaves and pods. This disease may be particularly bad in wet summers. **Control:** spray with Benomyl or Thiophanate-methyl. **6** Flower dropping (see Physiological disorders). **Symptoms:** few pods, as flowers keep falling off. Variable weather causing irregular temperatures and supplies of water will lead to flower dropping. **Control:** grow beans in a reasonably sheltered position to avoid cold winds; mulch with grass cuttings to keep even moisture at roots, but *do not* use cuttings from grass that has been treated with hormone weedkiller. **7** Foot rot, caused by *Fusarium* species of fungi (see Foot rot). May attack young and established plants. **Symptoms:** wilting and yellowing of foliage, followed by wilting of plants and death. **Control:** in future, use seed dressings Captan or Thiram (see Seed dressings); practise rotation and sterilize soil chemically (see Rotation, Soil sterilization). **8** Halo blight *Pseudomonas phaseolicola* (see Bacterial diseases). **Symptoms:** yellow circles on stems and leaves, which wilt, and blisters on seeds. **Control:** copper sprays may save a growing crop; little else can be done apart from obtaining clean seed. **9** Red spider mite *Tetranychus urticae* (see Mites). May attack established plants. **Symptoms:** leaves turning pale; covering of underside of leaves with minute red spider mites and their webs. **Control:** spray regularly with plain water, or if this fails, with Derris, Dimethoate or Malathion. **10** Rust, caused by the fungus *Uromyces phaseolorum*. May attack established plants. **Symptoms:** reddish pustules on the leaves. **Control:** unnecessary. **11** Sclerotinia, caused by the fungus *Sclerotinia sclerotiorum*. May attack young or established plants. **Symptoms:** white mouldy areas, and later black spores on stems at ground level. **Control:** practise rotation and sterilize soil chemically (see Rotation, Soil sterilization). **12** Slugs (see Slugs and snails). **13** Virus, Mosaic virus (scc Virus diseases). May attack established plants. **Symptoms:** light and dark mottling on the leaves. **Control:** use virus-free seed.

Beech *Fagus sylvatica* and varieties and other *Fagus* species and varieties. Hardy deciduous trees. Height: large. Foliage: beautiful, especially in spring and when coloured in autumn. Some varieties have copper or purple foliage, or ferny foliage. Use: planted singly or in groups, avenues, hedges.

Pests and Diseases. 1 Aphis, Beech aphid *Phyllaphis fagi* (see Aphis). May attack young and established plants. **Symptoms:** unsightly aphids and honeydew on the leaves, especially in hedges (see Honeydew). **Control:** spray with insecticides to reduce aphids, if this is practical. **2** Beech coccus, Beech scale *Cryptococcus fagi* (see Scale insects). May attack young and established plants. **Symptoms:** white, waxy, woolly material on trunks and branches. This is a widespread pest of beech trees in woods and does little harm on healthy trees. It will however increase rapidly on trees that are unhealthy or are growing in poor conditions. **Control:** cut down and burn unhealthy trees; spray trees in winter with tar oil; if possible, scrub trunks and large branches with a paraffin/soap emulsion. **3** Bracket fungus (see 6 below Heart rot). **4** Damping off or Wilt, caused by the fungus *Phytophthora fagi* (see Damping off). May attack seedlings. **Symptoms:** dark blotches covered with white zones of mould on the stems. **Control:** spray with liquid copper or Benomyl or commercially with Aaterra, at the first sign of attack. **5** Gas poisoning. A common cause of death of beech hedges or trees, where town gas poisons and natural gas asphyxiate. **Symptoms:** poor growth and death of hedges or trees; a smell of gas may be noticeable, especially if the ground is opened up. Check with the Gas Board. **Control:** for natural gas damage, oxygen can be injected. **6** Heart rot or Bracket fungus, caused by the fungus *Ganoderma applanatum*. May attack established plants. **Symptoms:** rotting of the centres of trees, the first signs being the formation of brown bracket bodies on the lower parts of trees. **Control:** if possible, cut out affected branches; if the main stem is attacked, little can be done apart from cutting down and destroying trees. **7** Honey fungus, caused by the fungus *Armillaria mellea* (see Armillaria mellea). May attack established plants. **Symptoms:** gradual death of branches and finally whole trees; clusters of white, scaly, gilled fungal growths on infected or dead trees; the bark can be easily pulled away, showing beneath white fans of fungal growth. This is a serious disease which affects many different kinds of trees, and does extensive damage to beech trees. **Control:** many authorities claim that there is no really effective control other than removal of infected trees and, if possible, treatment of the ground with Basamid or formaldehyde. In recent years, phenolic emulsion, containing a specific amount of phenol, has been developed for the treatment of affected trees. **8** Root rot, caused by a *Phytophthora* species of fungus. May attack established plants. **Symptoms:** the fungus attacks the roots, then enters the trunk and branches, causing wilting and death. **Control:** remove and burn infected trees; improve drainage and general growing conditions, to safeguard trees from attack. **9** Scale (see 2 above Beech coccus or Beech scale). **10** Wilt (see 4 above Damping off or Wilt).

Beetroot   Varieties derived from *Beta vulgaris*, the wild beetroot. Not fully hardy, herbaceous perennial with thickened roots.

Annual. Height: small. Roots: cylindrical, globular or long, and thick. Colours: red, white or yellow. Use: food crop, grown for its roots.

Pests and Diseases. 1 Blackfly *Aphis fabae* (see Aphis). May attack at all stages. **Symptoms:** distortion of leaves; aphids visible on plants. **Control:** spray with Malathion. **2** Black Leg, caused by the following fungi: (a) an *Aphanomyces* species (b) *Pleospora betae* (c) a *Pythium* species. **Symptoms:** (a) dry rot of roots (b) if sown too early, seedlings may be attacked and may never fully recover; when the crown of plants are attacked, dry rot follows; this disease is seed-borne (c) damping off of seedlings (see Damping off); spores of this disease may be present in the soil. It is difficult to establish which of these three diseases is present. **Control:** as a precaution, use mercury-based or Thiram seed dressings (see Seed dressings); practise rotation (see Rotation). **3** Crown gall *Agrobacterium tumefaciens* (see Bacterial diseases). May attack established plants. May also attack other different plants. **Symptoms:** malformed roots. **Control:** not important. **4** Downy mildew caused by the fungus *Peronospora schachtii* (see Mildews). **Symptoms:** downy growth on the leaves; curling of leaf edges. **Control:** spray with copper fungicide, Maneb, Thiram or Zineb. **5** Flea beetle (see 8 below Mangold flea beetle). **6** Heart rot (see Physiological disorders). **Symptoms:** brown leaves; white rings in the centres of the roots, spoiling the quality. It is best to consult the local horticultural adviser if this condition is suspected. **Control:** Borax powder at the rate of 8–17 g per square metre ($\frac{1}{4}$–$\frac{1}{2}$ oz per square yard) is recommended, but must be used correctly. **7** Leaf spot, caused by the fungus *Cercospora beticola*. May attack established plants. **Symptoms:** grey spots with red edges on the leaves; centres of spots often drop out, giving a shot hole appearance. In bad attacks, the top leaves may be destroyed, but new leaves will grow up to replace them. This disease is unsightly rather than really damaging. **Control:** remove and burn diseased leaves regularly; spray with a copper fungicide, Maneb, Thiram or Zineb. **8** Mangold flea beetle *Chaetocnema concinna*. Attacks at all stages. **Symptoms:** very active beetles on plants, biting pieces out of the leaves. Other plants may be attacked. **Control:** scatter Carbaryl or gamma-HCH dust to destroy beetles. **9** Mangold fly *Pegomya hyoscyami*. May attack at all stages. **Symptoms:** eggs laid on undersides of leaves in batches; blisters caused by small larvae which tunnel between upper and lower layers of tissue. It is difficult to control once larvae are present in the leaves. **Control:** spray seedlings to discourage adults; spray with gamma-HCH or Malathion when leaves are affected; top dress plants with nitrate of soda to produce a more leafy growth. **10** Rust, caused by the fungus *Uromyces betae* (see Rust). May attack established plants. **Symptoms:** reddish pustules on leaves. **Control:** unnecessary, seldom causes serious damage. **11** Scab, caused by a *Streptomyces* species of fungus. May attack established plants. **Symptoms:** scabs on the roots, disfiguring but not harmful. This disease thrives in alkaline soils. **Control:** work in a little peat along the rows, especially in limy soils. **12** Slugs (see Slugs and snails). **13** Violet root rot, caused by the fungus *Helicobasidium purpureum*. May attack established plants. **Symptoms:** yellowing and drooping of foliage; violet fungal growth on roots. Affects a number of different plants. **Control:** practise long rotation (see Rotation). **14** Virus, Mosaic virus (see Virus diseases). May attack young and established plants. Disease spread by aphids. **Symptoms:** mottling of leaves. **Control:** spray with Malathion to destroy aphids; keep down weeds, as certain weeds may also be affected.

Begonia This genus (see Genus) contains a large number of species, divided into groups according to habit of growth. There are many varieties and hybrids (see Hybrid). Greenhouse, half-hardy, herbaceous or woody perennials, which may have bulbs, rhizomes or tubers, or may be fibrous-rooted. Height: small to medium, climbing, trailing. Flowering time: varies according to species and method of growing. Flower colours: apricot, pink, red, white, yellow or parti-coloured. Leaves: fleshy, and show great variation according to species; may be large, small, hairy, smooth or wrinkled and differently shaped, coloured and variegated. Use: greenhouse border and pot plants, indoor pot plant, summer bedding, containers in the open, window boxes.

Pests and Diseases. 1 Abnormal growth (see Physiological disorders). May occur at all stages. **Symptoms:** unbalanced growth; misshapen leaves, red or brown-edged, blackened or bleached. A number of factors may be responsible for the different symptoms: excess nitrogen, causing gross, unbalanced foliage; potash deficiency, causing discoloured leaf margins; exposure to strong sunlight, causing abnormal growth and blackening of the leaves. **2** Bacterial blight *Xanthomonas begoniae* (see Bacterial diseases). Affects winter-flowering varieties. May attack established plants. **Symptoms:** brown spots with translucent margins scattered over the leaves; leaves may fall prematurely; stems may be attacked and plants may die; if leaf stems of infected plants are broken, a yellow slime may ooze out. **Control:** destroy infected plants; avoid overhead watering if this disease is present; keep pots well spaced out; clean benches thoroughly; use clean containers; take cuttings from healthy plants only. **3** Broad mite *Polyphagotarsonemus latus*. May attack established plants. **Symptoms:** puckering and turning down of young leaves, which may also be brittle, caused by adults feeding; adults, light brown/grey in colour, visible in folds of the leaves. **Control:** spray with Dicofol repeatedly, or dust lightly with flowers of sulphur weekly. **4** Glasshouse whitefly *Trialeurodes vaporariorum* (see Whitefly). May attack at all stages. **Symptoms:** plants badly affected, with distorted, discoloured leaves and poor flowers. **Control:** spray regularly with suitable insecticides. **5** Greedy scale *Hemiberlesia rapax* (see Scale insects). May attack at all stages. **Symptoms:** weakening and general poor growth; small scales, each with a distinct cone, on leaves and stems. **Control** (see 7 below Hemispherical scale). **6** Grey mould, caused by the fungus *Botrytis cinerea* (see Botrytis). May attack young and established plants. **Symptoms:** grey/white mould on leaves, stems and flowers; poor growth. **Control:** improve growing conditions by making sure ventilation is adequate and by avoiding splashing water over the plants. **7** Hemispherical scale *Saissetia coffeae*. May attack at all stages. This reddish-brown, highly convex scale insect feeds on stems and larger leaf veins. **Symptoms:** copious honeydew on plants (see Honeydew); stunted growth. Begonias are very sensitive to pesticide treatment. **Control:** best achieved by careful removal of scales, followed by spot treatments with Nicotine. **8** Mites (see 3 above Broad mite). **9** Powdery mildew, caused by the fungus *Oidium begoniae* (see Mildew). May attack established plants. **Symptoms:** white, powdery mould on the leaves and stems; poor growth. Some begonias tend to be slightly damaged by sprays. **Control:** spray with a suitable fungicide e.g. Thiram; avoid drying out of compost. **10** Root and stem rots (see Root rot, Stem rot), caused by various species of fungi. May attack at all stages. **Symptoms:** rotting of roots and stems and death of plants. May be particularly bad at propagation stage. **Control:** use sterilized compost and clean containers; avoid excess humidity; spray with a suitable fungicide, as some begonias tend to be slightly damaged by sprays. **11** Scale insects (see 5 above Greedy scale, 7 above Hemispherical scale). **12** Thrips (see Thrips). **13** Vine weevil *Otiorhynchus sulcatus*. May attack at all stages, including dormant corms. Damage is caused by the yellow C-shaped grubs, which excavate large cavities in the fleshy underground parts of the plants and tunnel into the stem bases. Damage is more frequent in the autumn. **Symptoms:** wilting of foliage. **Control:** the adult weevil cannot fly, so apply banding grease as used for fruit trees, in rings around the pots to provide an effective barrier to the adult when it tries to lay eggs on the plants; incorporate gamma-HCH into the potting compost to prevent attack, but no insecticide will kill the larger grubs. **14** Virus (see Virus diseases). (a) Mosaic, caused by cucumber mosaic virus. **Symptoms:** yellow areas on the leaves, with brown spotting between the veins. **Control:** destroy infected plants; use insecticides to control aphids which spread this virus; cleanse and fumigate the greenhouse when empty. (b) Ringspot, caused by tomato spotted wilt virus. **Symptoms:** ring spots on the leaves; poor growth. **Control:** destroy infected plants; spray with insecticides to control thrips which spread this virus; cleanse and fumigate the greenhouse when empty. **15** Weevil (see 13 above Vine weevil). **16** Whitefly (see 4 above Glasshouse whitefly).

Berberis *Berberis* species and varieties. Mainly hardy, deciduous and evergreen shrubs. Height: dwarf to medium. Flowering time: spring. Flower colours: orange or yellow. Stems: spiny. Berries: ornamental in autumn, black, purple or red. Use: planted singly or in groups, borders, informal hedges.

Pests and Diseases. 1 Leaf spot caused by the fungus *Pseudomonas berberidis* (see Leaf spot, Stem rot). **Symptoms:** spotting on leaves. **Control:** none needed, as disease is not serious. **2** Rust, caused by the fungus *Puccinia graminis* (see Rust). **Symptoms:** rust-coloured pustules on leaves, which cause distortion. **Control:** no long term control, but spray with liquid copper, Thiram or Zineb or commercially with Mancozeb. Most evergreen *berberis* species and varieties are resistant.

Betula (see Birch)

Bigeneric hybrid (see Hybrid)

Biological control Biological control is the rule rather than the exception in nature. Those insects and mites which damage crops represent only the tip of the iceberg of pests. Most potentially damaging animals are kept under control by their attendant natural enemies such as predators, parasites and diseases. The indiscriminate use of chemical insecticides has frequently caused more harm than good, not only because of their side effects on wildlife, but also by killing natural enemies of plant-feeding insects and mites, allowing these animals to increase in numbers and become pests. Unfortunately, some of the better known examples of biological control agents are not efficient enough to prevent damage, and insecticides still have many important uses. Ladybirds will eventually halt aphid attacks but usually not before the harm has been done. Birds which feed on grubs in lawns may cause more damage than the grubs themselves if the turf is pulled up during the hunt. The glasshouse provides an ideal environment in which to put biological control into practice, but one of the main obstacles to its success is that natural enemies are normally spread too thinly in nature to give complete control. Under glass, predators and parasites can be introduced and confined in large numbers. Three biological control agents are available to amateur growers. These are *Encarsia formosa* a small parasitic wasp which attacks glasshouse whitefly, *Phytoseiulus persimilis* the Chilean predatory mite which feeds on glasshouse red spider mite, and *Bacillus thuringiensis* a bacterium which causes a disease in caterpillars. In addition, the ladybird *Cryptolaemus montrouzieri* which is a predator of mealy-bugs, is being studied with a view to commercial production, as are numerous parasites and fungal diseases of aphids. Certain points should be borne in mind when attempting biological control of glasshouse pests. 1 The natural enemies will attack only particular pests. They will die out once control is achieved without themselves causing damage. 2 Natural enemies are often more sensitive to pesticides than are the pests. Having started biological control, it is essential to reduce the use of pesticides, choosing only those suggested by the supplier of natural enemies. 3 With the exception of diseases, natural enemies cannot be expected to control large numbers of pests. The natural enemies, especially *Encarsia*, must be introduced soon after the first sighting of the pest each year. 4 The remains of pests on ornamentals may still disfigure the plants long after natural enemies have given good control. The technique is therefore more suitable for edible crops, as it has the advantage of not putting crops at the risk of being tainted with insecticide. 5

Many pests, particularly whitefly and spider mites have developed resistance to pesticides. In the future, biological control may become the only practicable method of preventing pest damage. 6 The activities of natural enemies such as *Encarsia* are of educational interest to adults and children alike.

Birch *Betula pendula*, Common, Silver or White birch and varieties and other *Betula* species and varieties. Hardy deciduous trees and shrubs. Height: trees; small to large, shrubs; medium to large. Flowering time: spring. Decorative catkins. Foliage and habit of growth: graceful and foliage colours to lemon-yellow in autumn. Bark: decorative, silvery-white in common birch, may be almost black, orange-red, reddish-brown or yellowish in other sorts. Use: planted singly or in groups, avenues.

Pests and Diseases. 1 Aphis, Birch aphids *Euceraphis* species (see Aphis). May attack at all stages. May be damaging to young nursery plants. **Symptoms:** aphids and honeydew on the leaves (see Honeydew); growth weakened. **Control:** use suitable insecticides (see Aphis). **2** Birch mite *Eriophyes* species (see Mites). May attack established plants. **Symptoms:** clusters of abnormally developed branches; eventual death of affected trees. **Control:** if attack is slight, cut branches back beyond points of infestation and dress cuts with tar; if attack is severe, no control is possible. **3** Heart rot and bracket fungus, caused by the fungus *Polyporus betulinus*. In Scotland the fungus *Fomes fomentarius* may also occur. May attack established plants. **Symptoms:** rotting of centres of branches, making them weak; flat, bracket-like bodies on the wood, buff-coloured with white margins. **Control:** none possible beyond cutting off the affected branches and sealing the wounds carefully with paint or mercury compound. **4** Rabbit *Cuniculus lagopus* (see Rabbit). May attack at all stages. **Symptoms:** bark is stripped about 45 cm (18 in) above the ground; in severe attacks, the tree may be completely girdled and die. **Control:** protect the lower part of the trunk by a collar of wire-netting turned outwards at the top; spray the repellant Anthraquinone over the lower part of the tree. **5** Rust, caused by the fungus *Melampsoridium betulinum* (see Rust). Attacks young and established plants and young trees in nurseries. **Symptoms:** masses of yellow spores on undersides of leaves, causing leaf fall. **Control:** spray with liquid copper, Thiram or Zineb or commercially with Mancozeb. This however is seldom practical for established trees. **6** Witches' brooms, caused by the fungus *Taphrina betulina*. May attack established plants. **Symptoms:** clusters of branches which look like huge birds' nests among the normal branches. These masses of dense, congested growth are stimulated by the fungus. Witches' brooms are unsightly rather than damaging. **Control:** cut them off with a sharp saw, which will be effective for some time. There are no effective long term control measures.

Birds In the garden these are pleasant and, on the whole, beneficial, since many kinds feed on a range of pests at all stages of development. Some birds, such as tits and robins, may do damage to fruit, but they are mainly insectivorous, and are of value in keeping down caterpillars, aphids, scale insects and other pests. They may also feed on weed seeds. Birds such as blackbirds and thrushes certainly do damage to fruits, and feed on ornamental berries, but they also keep down cutworms, slugs, snails and other soil pests and various insects. Sometimes birds may pull out young seedlings, or destroy buds, leaves and flowers. Woodpigeons and house sparrows, among others, can be destructive in this way.

Control: if on a small scale, use netting and if on a larger scale, use bird scarers or chemical repellents; take simple protective measures such as stretching black thread over seedling rows.

Blackberry The name blackberry is used for a number of different species and varieties of *Rubus*. The wild blackberry or bramble is *Rubus lacinatus*. Selected forms of different species and cultivated varieties are grown. There are a number of hybrids of blackberries and related berries, known by the general name of hybrid berries (see Hybrid). There are also varieties of these. Hardy deciduous shrubs. Height: spreading. Trailing habit of growth, with strong arched branches. Stems: spiny or smooth. Flowering time: spring. Fruit: orange, red, red-black or black. Use: food crop grown for the fruit, trained to supports. For **Pests and Diseases**, see Raspberry.

Blackcurrant Varieties derived from the common blackcurrant *Ribes nigrum*. Hardy deciduous shrubs. Height: medium. Flowering time: spring. Leaves: strong smelling. Fruit: black. Use: food crop, grown for the fruit. See also Red currants.

Pests and Diseases. 1 American gooseberry mildew, caused by the fungus *Sphaerotheca mors-uvae*. May attack established plants. **Symptoms:** shoots and fruits coated with white mould. This is principally a disease of gooseberries but may also attack blackcurrants. **Control:** spray with Benomyl, Dinocap or Thiophanate-methyl when the flower clusters look like a small bunch of grapes; spray with tar oil winter wash in the dormant season, including ground round the bushes. **2** Aphis, Blackcurrant aphid *Cryptomyzus galeopsidis*, Currant and sowthistle aphid *Hyperomyzus lactucae*, Red currant blister aphid *Cryptomyzus ribis* (see Aphis). May attack at all stages. **Symptoms:** marking and mottling of the leaves; conspicuous blisters (red currant blister aphid) on the leaves; weakening of growth; aphids and honeydew on the plants (see Honeydew). **Control;** spray with Dichlorvos, Derris, Dimethoate, or Malathion; as aphids may build up a resistance to particular chemicals, use varied sprays. **3** Big bud (see 4 Blackcurrant gall mite). **4** Blackcurrant gall mite *Cecidophyopsis ribis*. May attack at all stages. **Symptoms:** minute mites infest the leaf scales in spring, later entering the buds which turn swollen and rounded; enlarged buds may be seen on the bushes in winter, before they finally wither and die; in spring and early summer, the mites are dispersed by wind, other insects or birds' feet, and carried to other blackcurrant bushes. A very serious pest of blackcurrants (see 20 below Reversion). Occasionally attacks red currants and gooseberries. **Control:** spray with lime sulphur as the flowers begin to open and repeat 3 weeks later; commercially, Endosulfan is recommended, but it is highly poisonous. **5** Blackcurrant leaf midge *Dasineura tetensi*. May attack at all stages. **Symptoms:** small white larvae attack tips of shoots, causing distortion and checking growth. **Control:** spray as the flowers are opening with Carbaryl or Dimethoate and repeat if necessary 3 weeks later. **6** Blackcurrant sawfly *Nematus olfaciens*. May attack at all stages. **Symptoms:** caterpillars, green, speckled with blue dots and with black heads, start feeding in April–May, usually working round the margins of the leaves at first; in severe attacks, they may defoliate whole bushes, causing very considerable damage. **Control:** as soon as an attack is seen, spray with Derris, Fenitrothion or Malathion; as there may be several generations in the season, it may be necessary to spray again; Azinphos-methyl is also recommended, but it is a dangerous chemical and not available to amateur gardeners. **7** Capsid, Common green capsid *Lygocoris pabulinus*. May attack young and established plants. **Symptoms:** active green insects feeding on leaves cause mottling and distortion. **Control:** spray with Dimethoate, Fenitrothion, gamma-HCH or Malathion. **8** Caterpillars (see 5 above Blackcurrant leaf midge, 6 above

Balckcurrant sawfly, 10 below Currant clearwing, 12 below Currant shoot borer, 23 below Tortrix moths, 25 below Winter moth). **9** Clay-coloured weevil *Otiorhynchus singularis*. May attack young and established plants. **Symptoms:** brown weevils hide under the leaves by day, coming out at night to feed on the shoots (see Weevil). **Control:** spray with Carbaryl at the first signs of attack, making sure to spray well round the base of plants and over the ground. **10** Currant clearwing *Synanthedon salmachus*. May attack established plants. **Symptoms:** wilting or dead shoots; wilting blossom; if shoots are slit open lengthways, white larvae may be seen inside; moths lay eggs in June and larvae tunnel inside the shoots. This is not a widespread pest. **Control:** if it has been troublesome, spray at the end of May with Fenitrothion, and repeat the spray 10–14 days later; commercial growers may use Azinphos-methyl after picking, but it is a dangerous chemical and not available to amateur gardeners. **11** Currant eelworm *Aphelenchoides ritzema-bosi*. **Symptoms:** microscopic eelworms feed in the buds and leaves, causing distortion and failure of buds to open in spring. This pest is not common, but may be troublesome after a wet summer. **Control:** only control possible for amateur gardeners is to cut out and burn infested shoots; the chemical recommended for control is Parathion, a highly poisonous material not available to amateurs. **12** Currant shoot borer *Lampronia capitella*. May attack established plants. **Symptoms:** small brown moths with yellow markings are on the wing in May and June; eggs are laid and larvae bore into buds and stems, then overwinter as chrysalids on the bark or in the crevices. This pest is difficult to control. **Control:** spray with tar oil wash in December. **13** Earwig *Forficula auricularia* (see Earwig). May attack fruit. May be a troublesome pest to commercial fruit processers, but unlikely to cause trouble for the amateur gardener. **Control:** if necessary, spray with Carbaryl. **14** Eelworm (see 11 above Currant eelworm). **15** Fruit tree red spider mite *Panonychus ulmi* (see Mites). May attack established plants. **Symptoms:** (see Apple 33 Fruit tree red spider mite). **Control:** best control is tar oil winter wash in the dormant season; otherwise, spray with Dimethoate, Malathion or other suitable acaricide. **16** Glasshouse red spider mite *Tetranychus urticae* (see Mites). May attack established plants. **Symptoms** (see Tomato 18 Glasshouse red spider mite). This mite overwinters as an adult in rubbish or soil at the base of the plants and not in the egg stage, which makes control difficult. **Control:** spray at fruit setting stage with Derris, Dichlorvos, Dimethoate or Malathion; apply a tar oil winter wash in the dormant season to give a certain degree of control. **17** Leaf spot, caused by the fungus *Pseudopeziza ribis.* May attack young and established plants. This is a troublesome disease of blackcurrants, especially in wet seasons. **Symptoms:** leaves show brown spotting and may fall early; growth is weakened. **Control:** spray as for 1 above American gooseberry mildew; apply Zineb which is also very effective if applied in April and repeated 3 or 4 times at 14-day intervals; use a copper fungicide after picking fruit. **18** Mildew (see 1 above American gooseberry mildew). **19** Red spider mite (see 15 above Fruit tree red spider mite, 16 above Glasshouse red spider mite). **20** Reversion, caused by virus infection (see Virus diseases). **Symptoms:** flower buds are reddish and the colour of the flowers seems darker; of more significance is the reduction in the number of veins on the leaves, causing them to alter shape, fewer than 5 lateral veins running from the midrib of affected leaves to each leaf edge; bushes change in appearance, having loose, leafy top growth, and narrow, pointed leaves; eventually no fruit is produced; reverted bushes become worthless. This disease is spread by the blackcurrant gall mite (see 4 above Blackcurrant gall mite) and possibly by other pests too. **Control:** spray against aphids, gall mites and other insects; do not take cuttings from infected bushes; destroy infected bushes; buy new certified stock (see Certified stock) and plant in fresh ground. **21** Rust, caused by the fungus *Cronartium ribicola* (see Rust). May attack established plants. **Symptoms:** leaf spotting similar to a leaf spot attack (see 17 above Leaf spot), but the spots are on the undersides of the leaves; leaves turn brown and fall early. **Control:** spray with Zineb in early April and repeat several times or a copper fungicide may be used. **22** Scale insect, Mussel scale *Lepidosaphes ulmi.* **Symptoms** (see Heath 5 Scale). **Control:** spray with tar oil wash in dormant season. **23** Tortrix moths. May attack established plants. **Symptoms** (see Apple 35 Fruit tree tortrix moth). **Control:** spray as first flowers open, and 3 weeks later, with Carbaryl or Fenitrothion. **24** Weevil (see 9 above Claycoloured weevil). **25** Winter moth *Operophtera brumata.* May attack established plants. **Symptoms** (see Apple 53 Winter moth). **Control:** spray as first flowers open, and 3 weeks later, with Carbaryl or Fenitrothion.

Blackfly   (see Aphis)

Blueberry Highbush and lowbush varieties. Highbush varieties *Vaccinium corymbosum* and *Vaccinium australe* (in cool temperate regions) and *Vaccinium ashei* (in warm temperate regions) are complex hybrids (see Hybrid) of *Vaccinium* species. Lowbush varieties are found in the wild although some have been produced through systematic breeding in recent years. Hardy deciduous shrubs. Height: dwarf to small. Flowering time: spring. Flower colours: white, cream, reddish. Fruit: blue-black, often with bloom. Use: food crop, grown for its fruit. In North America, many insects (nearly 300 species have been recorded) attack the blueberry. Because of the wide range of environments, different insects are important in different areas. It is therefore advisable to consult the local State Extension Service for information about pests and diseases and their control. The most troublesome pests in Britain are grazing animals (for example rabbits and horses) which eat young growth and bark, and birds (mostly blackbirds and thrushes) which eat the fruit. Plants can be protected by netting. The few infestations of aphids and caterpillars (tortrix moth, winter moth) are easily controlled by appropriate sprays when symptoms are observed but insect pests have not become important. Diseases common in Britain are listed below.

Pests and Diseases. 1 Botrytis, caused by the fungus *Botrytis cinerea*. May affect flowers but mostly affects the tips of shoots. **Symptoms:** varying degrees of dieback on tips of shoots. **Control:** if the disease proves to be troublesome, apply 3 sprays of Dichlofluanid at 7 to 10-day intervals during blossom. **2** Mummy berry, caused by the fungus *Monilinia vaccinii—corymbosi*. This disease attacks in the wild but is not yet serious in cultivation. **Symptoms:** seeds abort and fruit mummifies to a pale tan colour. **Control:** no practical control. **3** Plant degeneration, caused by infection by virus diseases. **Symptoms:** loss of vigour; yellow or red discolouration of leaves, sometimes showing patterns; malformation and reduction in size of leaves. **Control:** root out and burn. **4** Root rot, caused by species of *Phytophthora*. Affects only single plants and can be damaging. **Control:** root out and burn. **5** Stem canker, caused by the fungus *Godronia cassandrae*. A serious disease. **Symptoms:** small red discolourations on the stems which spread, turn dark brown with reddish margins, and girdle the stem and kill it. **Control:** grow the less susceptible varieties; cut out and burn as much affected wood as possible; Captofol sprayed on regularly throughout the growing season has been effective in the USA.

Botrytis   There are several species of this fungus, but the most familiar, widespread and serious of them is *Botrytis cinerea* which

Botrytis cinerea on 1 Strawberry fruit and 2 Tomato stem

can attack a very wide range of plants, both in the open and in the greenhouse. Names given to *Botrytis cinerea* may vary according to the plant affected, but botrytis, grey mould, and mildew, are commonly used. Typical symptoms are soft rotting of leaves and stems, which later become covered with soft, smoky grey fungal growth. Spots may occur on leaves, flowers and fruit, which later rot.

Control: remove and destroy all diseased, dying or dead material, and grow plants as well as possible; avoid overcrowding in the greenhouse; use ventilation and warmth to create a buoyant atmosphere; in the greenhouse, spray with Benomyl or Dichlofluanid; in the open, use Captan, Thiram or Benomyl.

Bramble   (see Blackberry)

Brassicas   The genus Brassica includes a number of species from which the following vegetables have been derived: broccoli, Brussels sprouts, cabbage, calabrese, cauliflower, radish, savoy, swede and turnip. (For descriptions, see under individual names.) Brassicas are all affected by the same pests and diseases.

Pests and Diseases. 1 Aphis, Cabbage aphis *Brevicoryne brassicae* (see Aphis). A widespread and troublesome pest of all the brassicas. May attack at all stages. **Symptoms:** grey-green, mealy or waxy aphids on plants; seedlings attacked are unfit for planting; distorted and dwarfed leaves on older plants; small black eggs, laid in the autumn on Brussels sprouts and broccoli, overwinter and hatch out in March and April. Winged aphids migrate to other brassicas and after several generations, egg laying starts again. **Control:** use proprietary insecticides following instructions carefully, especially noting time advised between use of spray and eating vegetables. Commercial growers use, among others, Dichlorvos, Dimethoate, Formothion, Malathion and Menazon. In gardens, materials such as Derris and Pyrethrum may be used. Nicotine is effective, but it is dangerous to use, and a suitable interval must be allowed between spraying with nicotine and eating vegetables. **2** Bacterial leafspot *Pseudomonas maculicola* (see Bacterial diseases). May attack established plants. **Symptoms:** small dark brown or purple spots on leaves. This disease is seldom seen except on brassicas grown for seed production. **Control:** destroy all affected plants. **3** Blackleg or Canker, caused by the fungus *Phoma lingam*. May attack young or established plants. First observed in the autumn months on winter brassicas. **Symptoms:** a black coating on stems, in bad cases encircling the stems; spores on stems, and these spread the disease. Broccoli are particularly affected by this disease. **Control:** ensure seed is clean. Commercially, seed is given a special hot water treatment or is soaked in Thiram. **4** Black ring spot (see 35 below Virus (a) Black ring spot virus). **5** Black rot *Xanthomonas campestris* (see Bacterial diseases). May attack at all stages. **Symptoms:** black rotting areas on plants, causing a yellowing of the leaves. **Control:** in future, sterilize seed-bed soil with Basamid (see Soil sterilization). **6** Cabbage aphis (see 1 above Aphis). **7** Cabbage gall weevil or turnip gall weevil *Ceutorrhynchus pleurostigma*. May attack young and established plants. **Symptoms:** in all cases, gall-like swellings, often similar to those caused by club root on the roots and necks of the plants, just below ground level; eggs, are laid in the spring, inside small holes drilled by the weevils; galls and stunted growth caused by larvae hatching out and eating the plant tissue. Cabbages and turnips mainly affected. **Control:** no really satisfactory method of control. HCH, HCH/Captan or HCH/ Thiram sprayed on after planting, or when plants are a few centimetres (an inch or two) tall, may be helpful. **8** Cabbage moth *Mamestra brassicae.* May attack young and established plants. Caterpillars attack most brassicas and also other plants, such as lettuce, onion and tomato. **Symptoms:** in almost all cases, the growing tips of plants are damaged. The large brown moths, with white markings on the first pair of wings, are mainly active at night from June onwards, laying globular white eggs in the centres of plants. The caterpillars are similar to those of the cabbage white butterflies, pale green at first, changing as they grow to darker green or greenish-brown. The main difference is that they are quite smooth, and not hairy or velvety. There may be a second generation in summer. The caterpillars are very destructive in feeding. **Control:** spray or dust with suitable chemicals, such as Carbaryl, Derris, Fenitrothion, Formothion or Trichlorphon; always follow directions very *carefully*, especially noting time advised between use of chemicals and eating vegetables. **9** Cabbage root fly *Delia brassicae.* May attack young and established plants. **Symptoms:** flies, resembling common house flies, appear about mid-April, or later if the weather is cold, after overwintering in the soil as pupae. Eggs are laid around the necks of the plants and after 6 days, small legless larvae emerge which eat into the roots and burrow into the stems, damaging and destroying many plants. In about 3 weeks, the larvae pupate and another generation follows. Sometimes there may be a third generation. Control measures are becoming less effective due to the rapid build-up of strains (see Strain) of root fly resistant to various chemicals. **Control:** choose from the chemicals HCH, Diazinon, Dimethoate, Bromophos and Trichlorphon, though these are not fully effective. Old-fashioned methods are becoming more popular again, such as the use of Naphthalene, or of 50–70 mm (2–3 in) felt discs placed round the necks of newly set out plants to prevent flies from laying eggs. There is experimental work being done on the use of plastic discs in place of the older felt discs. **10** Cabbage stem weevil *Ceutorrhynchus quadridens* (see Weevils). May attack young and established plants. **Symptoms:** adult weevils, about 4 mm ($\frac{1}{6}$ in) long, rounded and covered with greyish scales, sometimes with a white spot in the middle of the back, on the plants in early summer. Eggs are laid in the leaf stalks, sometimes causing swellings at the points of entry. The larvae tunnel down through the leaves and stems towards the plant bases, causing stunting and making transplants brittle and difficult to handle. **Control:** in future, if this pest has been troublesome, use a gamma-HCH seed dressing, but this may need to be followed by a spray of gamma-HCH to the seedlings when the true leaves have begun to develop. **11** Cabbage white butterflies, Green-veined white *Pieris napi*, Large white *Pieris brassicae*, Small white *Pieris rapae*. May attack

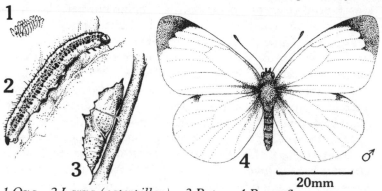

1 Ova   2 Larva (caterpillar)   3 Pupa   4 Butterfly

young and established plants. **Symptoms:** unless checked, the caterpillars will destroy plants completely. They skeletonize leaves, eat away broccoli and cauliflower curds and foul plants with their excreta. Eggs are laid from April onwards on the underside of leaves. The green, hairy or velvety caterpillars feed freely, then change to chrysalids and shelter in hedges and other places. There may be second and third generations in the summer. **Control:** spray or dust with suitable chemicals, such as Carbaryl, Derris, Fenitrothion, Formothion or Trichlorphon; always follow directions *very carefully*, especially noting time advised between use of chemicals and eating vegetables. **12** Cabbage whitefly *Aleurodes proletella* (see Whitefly). May attack established plants. **Symptoms:** white adult flies visible; flat, brownish scale-like larvae attached to underside of leaves; honeydew and sooty moulds on leaves, causing yellowing and poor growth (see Honeydew). This pest may be troublesome in warmer areas, especially on Brussels sprouts. **Control:** control measures for aphids are likely to be effective; remove and burn affected leaves. **13** Canker (see 3 above). **14** Caterpillars (see 8 above Cabbage moth, 11 above Cabbage white butterflies, 19 below Diamond back moth). **15** Club root or Finger and toe, caused by the fungus *Plasmodiophora brassicae*. May attack at all stages. A serious disease. The organism responsible for this disease is a slime mould or fungus which invades the root tissue. **Symptoms:** distortion of root tissue; inability of roots to take up water and nutrients; in the seed bed, poor growth of seedlings and, when lifted, swellings seen on the roots; wilting of young plants in bright sunshine, especially in cabbages and cauliflowers. As the season progresses, plants may be developing normally but if their roots are examined, their malformed or clubbed appearance can be clearly seen. These roots degenerate into a rotten, ill-smelling mass later in the season, especially in the case of turnips. Spores of the fungus are released into the soil, where they remain for a considerable period. Young plants raised in the greenhouse may be infected with club root. Seedlings may show signs of distress so examine young plants before transplanting and destroy infected ones. Clean containers and sterilized composts should always be used for young plants. **Control:** lime ground for seed beds and growing areas to give a pH of about 7 in spring; puddle roots in a paste of calomel dust before planting; practise rotation (see Rotation); in extreme cases, avoid growing any brassicas for a number of years at least; in gardens, sterilize the seed bed and/or the growing area with formaldehyde (see Soil sterilization) or Basamid (Dazomet), a metham sodium based soil sterilant in powdered form, especially for seed beds. Directions must be carefully followed. Alternatively, calcium carbide (Spent) may be used, if available, but it is a very unpleasant material to handle. **16** Cutworms (see Cutworms). **17** Damping off or wire stem, caused principally by the fungus *Rhizoctonia solani*, and other species of fungi (see Damping off). May attack at all stages. **Symptoms:** collapse at soil level of seedlings, especially those raised in the greenhouse; darkening and thinning at soil level of stems of older plants, when the disease is called wire stem. **Control:** water seedlings with Cheshunt compound, Quintozene or Thiram which are reasonably effective; it is best however to use sterilized composts or to sterilize soil in seed beds (see Soil sterilization) with Basamid or commercially with metham sodium *before* sowing seed, following all directions very carefully, and allowing the recommended time to elapse before sowing. **18** Dark leaf spot, caused by the fungus *Alternaria brassicicola*. May attack established plants. **Symptoms:** roundish grey areas on the leaves. **Control:** seldom really damaging, but spray with a copper fungicide if necessary. **19** Diamondback moth *Plutella* *xylostella*. **Symptoms:** small 6 mm ($\frac{1}{4}$ in) moths appear in May and June and lay eggs which hatch out light green caterpillars which feed on the leaves. In 3 weeks they spin cocoons and pupate. Later adults emerge, more eggs are laid and more caterpillars hatch out. This pest overwinters in the soil. It is troublesome only occasionally, in hot, dry seasons, when most brassicas, especially turnips and swedes, may be attacked. **Control:** use Carbaryl, Derris or HCH dust. **20** Downy mildew caused by the fungus *Peronospora parasitica* (see Mildew). May attack at all stages. **Symptoms:** downy growth on leaves, followed by distortion of leaves and stems. Many different brassicas are attacked, especially those grown in the greenhouse at seedling stage. **Control:** spray with Mancozeb/Zineb, Thiram or Zineb, or with Bordeaux mixture which, though old-fashioned, is still effective. **21** Finger and toe (see 15 above Club root). **22** Flea beetles (see Flea beetles). **23** Grey mould, caused by the fungus *Botrytis cinerea* (see Botrytis). May attack at all stages. **Symptoms:** areas of soft, mouldy rot on the plants; seedlings may be destroyed. This disease attacks a wide range of plants in wet, humid conditions, and is especially damaging when plant growth is soft. **Control:** excellent control is possible by the use of Benomyl or Dichlofluanid, especially on seedlings, provided strains (see Strains) of botrytis resistant to Benomyl are not involved. **24** Leaf spot (see 2 above Bacterial leaf spot, 18 above Dark leaf spot, 25 below Light leaf spot, 29 below Ring spot, 32 below Soft spot, 35 below Virus). **25** Light leaf spot, caused by the fungus *Gloeosporium concentricum*. May attack established plants. **Symptoms:** round purple patches on older leaves. **Control:** none necessary. **26** Mildew, caused by the fungus *Erysiphe polygoni* (see Mildew). May attack at all stages. **Symptoms:** areas of powdery growth on the leaves. May attack all brassicas during dry weather, but not usually serious. **Control:** use one or other of the mildew sprays, such as Benomyl or Dinocap, but spraying is only necessary in extreme cases. **27** Mosaic (see 35 below Virus). **28** Mustard beetle *Phaedon cochleariae*. May attack at all stages. **Symptoms:** adult beetles are about 3 mm ($\frac{1}{8}$ in) long and shiny dark blue in colour. They overwinter in hedge bottoms. Eggs are laid in May or June and the small yellow larvae feed on various brassicas. A second brood appears in summer. **Control:** dust with Carbaryl, Derris or other insecticidal powder. **29** Ring spot, caused by the fungus *Mycosphaerella brassicicola*. May attack established plants.

Symptoms: brown spots on the larger leaves, causing them to turn yellow and die. **Control:** difficult to control, but high potash feeding is helpful. **30** Root knot, caused by a parasitic eelworm *Heterodera radiciola* (see Eelworms). May attack at all stages. **Symptoms:** similar but smaller clubbing of the roots as compared to club root (see 15 above Club root). **Control:** practise rotation (see Rotation). Difficult to control otherwise due to its wide host range (see Host). Not serious in Britain but frequently recorded in the USA. **31** Slugs (see Slugs and snails). **32** Soft spot *Erwinia carotovora* (see Bacterial diseases). May attack young or established plants. **Symptoms:** mushy black areas on plants. All brassicas, but cauliflowers in particular, may be attacked. **Control:** good drainage of land along with adequate potash fertilizer dressings may be helpful; application of boron may also be helpful, but take advice from the local horticultural adviser if boron is to be used. **33** Turnip fly (see Flea beetles). **34** Turnip gall weevil (see 7 above Cabbage gall weevil). **35** Virus (a) Black ring spot (b) Mosaic virus (see Virus diseases). (a) Black ring spot virus. **Symptoms:** small black rings on the leaves. **Control:** seldom very damaging, so no control usually needed. (b) Mosaic virus. **Symptoms:** veins of leaves white, dull and mottled, eventually dropping off, leaving centre leaves only. **Control:** remove *badly* affected plants. Stimulate less seriously affected plants to grow by topdressing, when suitable, with extra nitrogen. **36** Weevils (see 7 above Cabbage gall weevil, 10 above Cabbage stem weevil). **37** Whiptail (see Physiological disorders). May attack established plants. **Symptoms:** leaf blades turn thin and narrow, only the midrib remaining. Due to lack of molybdenum in the soil. **Control:** apply sodium molybdate at 10 g per 16 litres (1 oz per 10 gallons) water and water this on to the plants. **38** White blister, caused by the fungus *Albugo candida*. May attack seedlings. **Symptoms:** smooth glossy patches on the underside of leaves on seedlings in the seed bed. **Control:** remove and burn affected plants. In future, sterilize seed bed with Basamid (see Soil sterilization). **39** Whitefly (see 12 above Cabbage whitefly). **40** Wilt (see 41 below Yellows). **41** Yellows, caused by the fungi *Fusarium conglutinans, Fusarium oxysporum*. May attack at all stages. **Symptoms:** yellow leaves. Affects all brassicas, but is mainly a disease of warm countries in temperatures above 20°C (68°F) and not usually a problem in Britain. **Control:** use specially treated seed.

Breeding   (see Plant breeding)

Broad bean   (see Bean, broad)

Broccoli   (see Brassicas, Cauliflower)

Brussels sprouts   Varieties derived from *Brassica oleracea bullata*. Hardy biennial, with stem woody at the base. Height: medium to large. Foliage: buds develop from base of stem upwards, making small, compact, leafy heads. Use: food crop, grown for the leafy foliage buds. For **Pests and Diseases**, see Brassicas.

Budding   A method of propagation used to produce new individuals quickly, and for various other reasons. A bud or eye is taken from the stem of a plant and inserted into the rooted stem of another plant, called the stock. There are various ways of doing this but it is an operation requiring skill and care.

Bulb mite   *Rhizoglyphus echinopus* usually attacks various bulb plants such as hyacinth, lily, narcissus, tulip and dahlia (stored tubers) after damage has been done by diseases or other pests or mechanically, or by incorrect storage. Swarms of the large, creamy-white, globular mites may be seen moving slowly on the bulb scales or on the lower leaves of growing plants. The mites increase rapidly on stored bulbs and tubers after planting, and in bad cases may destroy them. **Control:** examine carefully all soft bulbs and obviously unhealthy plants and destroy them if badly infected. Commercial bulb growers use hot water treatment for bulbs, 4 hours at 43°C (110°F) (see Hot water treatment). This is *not* suitable for tulips intended for forcing. See Bulb scale mite, Mites.

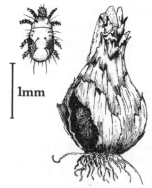

Rhizoglyphus echinopus

1mm

Bulb scale mite   *Steneotarsonemus laticeps*, unlike the bulb mite (see Bulb mite) is microscopic and attacks bulbs directly, particularly amaryllis and narcissus, especially in greenhouse conditions as the mites breed in warmth only. The mites live between the bulb scales. Stored bulbs infested by mites appear soft, dry and with the scales pressed closely together, owing to loss of moisture caused by the mites feeding. If a bulb is cut across just below the neck, small pieces of brown tissue will be seen at the tips of the scales. From the tips, discoloured areas run down the scales, indicating where the mites are feeding. On growing plants, the leaves seem abnormally bright green, later flecked with yellow, and distorted. Flower stems show a nicked, saw edge effect, and flowers may be badly distorted, or destroyed.

Control: examine, and destroy if necessary, all suspected bulbs. Commercial growers use hot water treatment, 4 hours at 43°C (110°F) or use Thionazin as a bulb dip (see Hot water treatment). See also Mites.

Busy lizzie *Impatiens sultani* and varieties and other *Impatiens* species. Hardy and greenhouse annuals, biennials or herbaceous perennials. Height: small to large. Flowering time: spring to autumn in the open, and in the glasshouse and indoors. Flower colours: orange, pink, red, white or parti-coloured. Use: hardy sorts in borders, by water, in woodland gardens; greenhouse sorts as greenhouse or indoor pot plants, or for summer bedding.

Pests and Diseases. 1 Leaf spot, caused by an *Ascochyta* species of fungus. **Symptoms:** spots on leaves and small black dots, which are the fruiting bodies of the fungus. **Control:** spray with Captan, if badly infected. **2** Sun scorch (see Physiological disorders). **Symptoms:** leaves browned and scorched round margins. **Control:** burning is caused by excess sunlight, especially if any water is lying on the leaves, so shade the greenhouse, or keep indoor plants away from direct sunlight, though not in too shaded a position.

Butterflies   (see Moths)

C

Cabbage   Varieties derived from *Brassica oleracea capitata*. Hardy biennial, with stem woody at the base. Height: small to medium. Foliage: leaves form a large, compact head, pointed or rounded, or a loose, leafy head. Use: food crop, grown for the leafy heads. For **Pests and Diseases**, see Brassicas.

Cactus   There is a genus *Cactus* (see Genus) in the *Cactaceae* family (see Family) but the name cactus is generally used for all the members of the family. It is a very large family and for convenience, is divided into various sections. Height: creeping to tall or climbing. Flowering time: throughout the year. Flowers: vary in size, form and colour, some are very beautiful. Spines: appearance enhanced by hairy, felt-like or woolly growths. Stems: interesting and decorative, enlarged, succulent, modified to conserve moisture, usually leafless although in one genus the plants are shrubs or small trees with large leaves, varied in size, shape and colour according to species. Use: indoor pot plants, greenhouse border and pot plants, climbers trained to greenhouse rafters, commercial glasshouse pot plants. Some species may be planted or kept outside during the summer and, under favourable conditions, some species may be grown in the open all year, though this is unusual.

Pests and Diseases. 1 Aphis, Mottled arum aphid *Aulacorthum circumflexum* (see Aphis). May attack at all stages. **Symptoms:** greenfly is always wingless and has a black U-shaped mark on its back; it breeds rapidly on poorly grown plants in low light and disfigures cacti by the accumulation of waste products and cast skins. **Control** (see Aphis). Note that gamma-HCH is not effective against this species. **2** Cactus root mealybug *Spilococcus cactearum* (see Mealybugs). Attacks young and established plants during the time when, in normal cultivation, water is being withheld. **Symptoms:** white powdery dust will be seen on the roots when re-potting, and closer inspection will reveal mealybugs. **Control:** apply a few crystals of paradichlorobenzene (sold as moth repellent) to the crock over the drainage hole when re-potting, and keep the soil moist for as long as possible. **3** Citrus mealybug *Planococcus citri*. May attack at all stages. **Symptoms:** greyish sunken areas at points where the mealybugs have fed. **Control:** spray with gamma-HCH or Nicotine. Other pesticides may cause damage and should be used cautiously. **4** Corky scab (see Physiological disorders). May occur on established plants. **Symptoms:** corky spots on the stems of many kinds of cacti, especially *Opuntia*; surface tissues crack and curl back and areas may be killed. This disorder is caused by unfavourable growing conditions, such as fluctuating temperatures, inadequate ventilation or unbalanced food supplies. **Control:** improve growing conditions; cut out badly-marked sections of plant. **5** False red spider mites *Brevipalpus* species (see Mites). May attack established plants. **Symptoms:** dry grey patches on the stems; fine webs stretched between spines. **Control:** use Dicofol at 7-day intervals. (*Not* for Epiphyllums.) **6** Glassiness (see Physiological disorders). **Symptoms:** areas on the stems with a water-soaked appearance. **Control** (see 4 above Corky scab). **7** Mealybugs (see Mealybugs, 2 above Cactus root mealybug, 3 above Citrus mealybug, 12 below Vine mealybug). **8** Mottled arum aphid (see 1 above Aphis). **9** Red spider (see 5 above False red spider mites). **10** Root and stem base rot, caused by *Phytophthora* species of fungi. May attack established plants. **Symptoms:** dark marks at the base of the stems. **Control:** cut out all diseased tissue completely; dust the cuts with flowers of sulphur. **11** Scale insects *Hemiberlesia rapax* and other species (see Scale insects). May attack established plants. **Symptoms:** weakening and general poor growth; small scales, each with a distinct cone, on the stems. **Control:** spray with Malathion, testing first with one plant for tolerance, or spray with Nicotine. **12** Vine mealybug *Pseudococcus obscurus* (see Mealybugs, 3 above Citrus mealybug).

Calabrese   (see Brassicas, Cauliflower)

Calceolaria   *Calceolaria* species and varieties. Half-hardy and greenhouse annuals, herbaceous perennials and shrubs. Height: small to medium, shrubs small. Flowering time: summer, autumn or winter according to variety. Flower colours: orange, red, yellow or parti-coloured. Use: greenhouse border, greenhouse pot plant, summer bedding, window boxes and outdoor containers. The herbaceous varieties with very showy, slipper-shaped flowers, are of particular value for greenhouse decoration. They are treated as biennials.

Pests and Diseases. 1 Glasshouse potato aphid *Aulacorthum solani*. May attack at all stages. **Symptoms:** distortion of leaf. **Control:** spray with Malathion but take care. **2** Glasshouse white-fly *Trialeurodes vaporariorum* (see Whitefly). May attack at all stages. **Symptoms:** leaves covered with whitefly and honeydew (see Honeydew). **Control:** spray with Diazinon or Malathion but take care with Malathion. **3** Root rot, caused by the fungi *Phytophthora* species and *Thielaviopsis basicola*. May attack at all stages. **Symptoms:** wilting and death, especially at the young stage. **Control:** water compost with Cheshunt compound, a copper fungicide or Zineb; use sterilized compost and clean containers. **4** Virus, Spotted wilt, caused by Tomato spotted wilt virus (see Virus diseases). May attack young and established plants.

Symptoms: stunted and distorted plants; pale irregular blotches and streaking on leaves; flowers reduced and distorted. **Control:** destroy affected plants; use insecticide to control thrips, which spread the virus; cleanse and fumigate greenhouse in winter, when empty.

Calendula   (see Marigold)

Callistephus   (see Aster)

Camellia   *Camellia japonica* and varieties and other *Camellia* species and varieties. Greenhouse (outdoors in mild areas), evergreen trees and shrubs. Height: trees, small; shrubs, small to medium. Flowering time: spring, summer. Flower colours: pink, red, white or parti-coloured. Use: planted singly or in groups outdoors in mild areas, greenhouse borders, greenhouse pot plants, outdoor containers in sheltered positions.

Pests and Diseases. 1 Bud shedding (see Physiological disorders). May occur on established plants. **Symptoms:** buds fall for no apparent reason. **Control:** avoid incorrect growing conditions, such as drought, over-watering, draughts, temperatures too high or too low, or excess nitrogen fertilizer. **2** Citrus mealybug (see 4 below Mealybug). **3** Leaf blotch, caused by the fungus *Pestalotia guepini*. May attack established plants. **Symptoms:** round brown or grey spots or blotches on the leaves. **Control:** remove and burn affected leaves. **4** Mealybug, Citrus mealybug *Planococcus citri* (see Mealybug). May attack at all stages. **Symptoms:** white, spiny mealybugs on the plants; honeydew and sooty moulds on the leaves (see Honeydew). **Control:** spray with Diazinon or gamma-HCH, but do not spray open blossoms or coloured flower buds. **5** Scale (see Scale insects). (a) Cushion scale *Chloropulvinaria floccifera*. May attack at all stages. **Symptoms:** dull brown insects, carrying large white masses of eggs, on the plants; plants disfigured. **Control:** take particular care to remove all traces of eggs before treating as for soft scale. (b) Soft scale *Coccus hesperidum* (see Rhododendron 24 Soft scale). (c) Vine weevil (see Weevils, Rhododendron 26 Vine weevil).

Capsicum   (see Pepper)

Capsid bugs   *Capsidae* form a large group of insects which feed by sucking plant sap. There are a number of different species, which may attack a wide range of plants. In appearance, capsid bugs are not unlike aphids. They may be greenish or yellowish in colour. Winged adults lay eggs on plants and the young capsid bugs feed, causing considerable damage to leaves, shoots, flowers or fruits, according to species and to the kinds of plants attacked. They leave these plants, move to different ones, and return later as winged adults, to begin egg laying again.

Control: spray bugs with Dimethoate or Fenitrothion, and repeat as necessary.

Carnation Modern glasshouse carnation varieties and the Border carnation have been developed from a series of different forms of *Dianthus caryophyllus*. The various pinks and Sweet William are derived from other *Dianthus* species. Some small *Dianthus* species and varieties are grown in rock gardens. Hardy, half-hardy and greenhouse annuals and herbaceous perennials. Height: dwarf to tall. Flowering time: throughout the year, according to sort and method of growing. Flower colours: cream, orange, pink, mauve, red, white, yellow or parti-coloured. Flowers: very fragrant. Use: greenhouse borders or pots for cut flowers, commercial glasshouse borders for cut flowers, commercial glasshouse pot plants, outdoor borders, annual borders, summer bedding, rock gardens. Note that the following pests and diseases mainly affect glasshouse carnations.

Pests and Diseases. 1 Anther smut, caused by the fungus *Ustilago violacea*. May attack established plants. **Symptoms:** stunted plants; numerous weak sideshoots; bushy habit; short flower buds; calyces split; flowers black and sooty. **Control:** remove affected plants before they flower; take cuttings from healthy stock. **2** Aphis, Glasshouse potato aphid *Aulacorthum solani* and Peach-potato aphid *Myzus persicae* (see Aphis). May attack at all stages. **Symptoms:** distorted leaves; aphids and honeydew on the plants (see Honeydew). **Control** (see Aphis). **3** Carnation fly *Delia cardui*. May attack in the autumn and winter. **Symptoms:** maggots feed in the stems and roots of carnations and pinks, occasionally causing leaf blisters. **Control:** take special care to prevent infested plants from being introduced into glasshouses, a period of quarantine being desirable; spray or dust plants with gamma-HCH at 3-week intervals from July to November; if leaf or stem damage is noted outdoors in September, spray with Diazinon or gamma-HCH. **4** Carnation spider mite *Tetranychus cinnabarinus*. May attack at all stages. **Symptoms:** tiny brick red or orange mites cover the plants with silken webs and cause whitish speckles on the leaves. **Control:** check introduced cuttings for damage and discard if in any doubt;

spray fortnightly with Dicofol, Dimethoate, Formothion or Malathion. **5** Carnation tortrix moth *Cacoecimorpha pronubana.* May attack established plants. **Symptoms:** young shoots and leaves spun together into tents by green caterpillars; flower buds attacked; black pupae visible on the foliage. **Control:** difficult, but spray repeatedly with gamma-HCH. **6** Caterpillars, various species. May attack established plants. **Symptoms:** caterpillars visible; holed and ragged leaves. **Control** (see Caterpillars). **7** Greasy blotch, caused by the fungus *Zygophiala jamaicensis.* May attack established plants. **Symptoms:** web-like patches on the leaves spread to give the surface an oily appearance. **Control:** reduce humidity in the greenhouse. **8** Grey mould, caused by the fungus *Botrytis cinerea.* May attack established plants. **Symptoms:** rotting of stems and wilting of shoots; fluffy fungal growth may be visible; petals rot and flowers may be destroyed. **Control** (see Botrytis). **9** Hollow flowers (see Physiological disorders). May occur on established plants. **Symptoms:** the number of petals is reduced and flowers appear to be single, with hollow centres. **Control:** maintain adequate watering; avoid excessively high temperatures by ventilation. **10** Leaf spot, caused by the fungus *Alternaria dianthi* and other species of fungi. May attack established plants. **Symptoms:** purple/brown spotting on leaves. **Control:** spray with Thiram or Zineb, or commercially with Mancozeb (see also 15 below Rust). **11** Mildew (see 12 below Powdery mildew). **12** Powdery mildew, caused by an *Oidium* species of fungus (see Mildew). May attack established plants. **Symptoms:** whitish powder on the leaves and occasionally on the calyces. **Control** (see Mildew). **13** Red spider (see 4 above Carnation spider mite). **14** Ring spot, caused by the fungus *Didymellina dianthi.* Attacks established plants. **Symptoms:** round or oval brown spots on leaves and stems, with dark powdery spores produced in concentric rings on the spots. **Control:** spray with Zineb; reduce humidity in the greenhouse. **15** Rust, caused by the fungus *Uromyces dianthi* (see Rust). May attack established plants. **Symptoms:** brown pustules on stems, leaves and on calyces. **Control:** destroy badly affected plants; on less badly affected plants, remove and destroy leaves with rust pustules; reduce humidity and overhead watering; spray commercially at 10 to 14-day intervals with Mancozeb/Zineb, or with Thiram or Zineb; take cuttings from healthy plants only. **16** Split calyx (see Physiological disorders). May occur on established plants. **Symptoms:** calyx splits down one side and petals spill out, the flower losing its compact shape. This condition is associated with the formation of additional whorls of petals, which cannot be contained by the calyx. **Control:** regulate day and night temperatures to avoid marked and rapid changes, which tend to stimulate production of these additional whorls of petals; use varieties not subject to split calyx. **17** Stem rots, caused by the fungi *Alternaria dianthi*, *Fusarium culmorum*, *Phytophthora* species, *Pythium* species and *Rhizoctonia solani*. Cuttings, newly rooted and newly planted plants may be attacked. **Symptoms:** pale or dark brown marking on stem bases, varying according to which disease is present. With *Alternaria*, small purple spots on the leaves may enlarge to brown patches with purple margins; with *Fusarium*, orange or pink pustules appear on the stems; with *Rhizoctonia*, stems break over at a node or just above soil level. **Control:** avoid over-watering; maintain a buoyant atmosphere; ensure thorough soil sterilization before planting (see Soil sterilization); to control *Rhizoctonia* commercially, incorporate Quintozene into the top 7.5–10 cm (3–4 in) of border soil before planting; spray plants, from which cuttings are to be taken, commercially with Captan or Maneb at 7-day intervals; select cuttings from healthy plants and spray or drench the cuttings with Captan during the rooting period; to control *Alternaria*, spray commercially with Mancozeb, or with Thiram or Zineb. **18** Thrips, Onion thrips *Thrips tabaci* (see Thrips). May attack at all stages. **Symptoms:** wilting and scarring of foliage. **Control:** spray regularly with Malathion or other suitable insecticide. Apart from the damage they do directly, thrips may spread virus diseases (see 19 below Virus). **19** Virus, various virus diseases (see Virus diseases). **Symptoms:** vary from slight or moderate mottling to more or less severe flecking with small dead spots on the leaves, and rings or streaks of red or yellow on the leaves. Most of the virus diseases are spread by handling and/or by aphids and thrips. **Control:** remove and destroy diseased plants; spray with insecticides to control aphids and thrips (see 2 above Aphis, 18 above Thrips). **20** Wilts, caused by a large number of species of fungi (see Wilts). May attack established plants and cause serious loss. (a) *Fusarium oxysporum*, *Fusarium* species *dianthi* and *Fusarium redolens*. **Symptoms:** leaves turn purplish-red, often on one side of the plant only; if stems are cut across, they will show brown centres; roots rot, so that plants can be easily pulled up;

plants die fairly quickly after infection. (b) *Phialophora cinerescens*. **Symptoms:** rapid wilting, often starting on one side of the plant; plants turn greyish-green, with some of the leaves becoming purplish-red; plants finally turn straw-coloured and die; if stems are cut across, they will show brown centres; roots do not rot, so the plants cannot be pulled up easily. (c) Slow wilt *Erwinia chrysanthemi* (see Bacterial diseases). May attack established plants. Wilting is gradual and plants may take 6–8 months to die. **Symptoms:** infected plants grow very slowly compared to healthy plants. **Control:** if wilts appear in pots, destroy plants, disinfect greenhouse and start again with sterilized compost, clean containers, and healthy plants; if wilts appear in a small area in the greenhouse border, immediately remove affected plants and those for about 90 cm (3 ft) around; remove soil to a depth of 60 cm (2 ft) and sterilize with a Formalin drench; fill in with clean soil; for more widespread infection, mark affected area and cultivate deeply after clearance of plants; sterilize with Formalin or Metham-sodium (Basamid) followed, if possible, by thorough steaming; disinfect the greenhouse; soil drenches with Benomyl will control *Fusarium* or *Phialophora* infection; take cuttings from healthy plants only, or buy from growers who have obtained indexed material (plants which have been propagated from parent material which has been produced under special conditions and carefully checked for the presence of disease); always use clean containers, disinfected if necessary and sterilized composts for cuttings, young plants and pot plants.

Carrot Varieties derived from *Daucus carota*. Biennial. Roots: swollen, fleshy, round, cylindrical or long and pointed. Colours: red or yellow. Use: food crop, grown for the edible sweet roots.

Pests and Diseases. 1 Aphis, Carrot root aphid *Pemphigus phenax*, Willow and carrot aphid *Cavariella aegopodii* and other *Aphis* species (see Aphis). May attack at all stages. **Symptoms:** bad distortion of growth above ground, although damage in some cases is also caused to roots; severe reduction in yield. The willow and carrot aphid may carry virus diseases, in which case there will be dwarfing of plants and further reduction of yield (see Virus diseases). There are several overlapping broods in the season, so controls must be repeated. There are a number of possible insecticides and, as the persistent use of any one chemical can result in strains (see Strain) of aphids resistant to this chemical, different insecticides should be used during the season. **Control:** use insecticides following directions *very carefully*, especially noting time advised between use of chemicals and eating vegetables. Derris is a safe chemical and may be used a few days before harvesting. Dimethoate and Malathion may be used, following directions. The older insecticide, Nicotine, is effective, but is *very dangerous*. For commercial growers the following are listed in *The Agricultural Chemicals Approval Scheme Booklet* (revised annually): Demephion, Demeton-S-methyl, Dimethoate, Disulfoton, Formothion, Malathion, Mevinphos, Phorate and Thiometon. Most of these are organo-phosphorus compounds and are therefore not advised for amateur use, except for Dimethoate or Malathion. **2** Bacterial soft rot (see 10 below Soft rot). **3** Black rot, caused by the fungus *Stemphylium radicinum*. A storage disease. **Symptoms:** blackening and rot from the crowns downwards on stored carrots. **Control:** remove and destroy affected carrots. In future take care to avoid any mechanical damage to the carrots during the growing season and when lifting, and store in suitable conditions. **4** Botrytis, caused by the fungus *Botrytis cinerea*. **Symptoms:** mouldy growth on the carrots while growing or when stored. Carrots may be badly attacked in wet seasons. **Control:** use Benomyl; avoid mechanical damage and remove any carrots showing mouldy growth in storage. **5** Carrot fly *Psila rosae*. A persistent pest which must be controlled. **Symptoms:** adult flies about 12 mm ($\frac{1}{2}$ in) in length and wing span, black with a brown head and yellow legs. Female flies appear in early May or June, according to region, guided by the smell of the carrots. Eggs are laid in clusters or singly in cracks in the soil close to the carrots where thinning has been carried out. The holes left are ideal for the eggs. The eggs hatch in about a week and the small larvae, which are pale yellow, feed on the smaller roots and then on the main roots, causing considerable damage. The leaves turn red and then yellow, and wilt. Growth is checked. After feeding for about 3 weeks the larvae pupate and the adult flies emerge in late July, giving rise to a second generation which also may attack carrots, but the effect is not so severe. **Control:** use chemicals from the wide range available, including both seed dressings and dusts (see Seed dressings); dust HCH with Captan or Thiram on to the seed *before* it is sown, which is probably the best and safest method for amateurs. Gamma-HCH dusts may be used from about mid-May to dust along rows every 10–14 days, and especially after

Brassicas **Cabbage root fly**
Delia brassicae, larvae eating
into root (see page 50)

Brassicas **Club root** caused
by the fungus *Plasmodiophora
brassicae*, distorted root tissue
(see page 51)

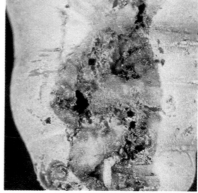

Carnation **Thrip** damage
caused by *Thrips tabaci*,
yellowing of calyx (see page
62)

Carrot fly *Psila rosae*, larvae
feeding on root (see page 64)

Chrysanthemum aphid
Macrosiphiniella sanborni on
stem (see page 71)

Chrysanthemum **Grey
mould** caused by the fungus
Botrytis cinerea, rotted flower
(see page 73)

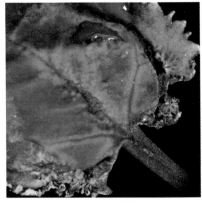

Chrysanthemum **Leaf miner**
Phytomyza syngenesiae, severe
attack, silvery-white, snake-
like markings on leaves,
caused by larvae tunnelling
inside (see page 72)

Cyclamen **Tarsonemid mite**
damage, caused by
Steneotarsonemus pallidus,
severe attack, outer leaves
curled upwards (see page 86)

thinning, in an effort to destroy eggs, or larvae as they hatch, and also to deter egg laying. A mulch of grass cuttings along the carrot rows may help, but *do not use* grass which has been treated with hormone weed killers. Thinning by cutting off tops rather than pulling out young plants, and firming of soil round plants after thinning, also helps to avoid leaving holes convenient for egg laying. The use of Naphthalene scattered along the rows to mask the smell of carrots, in addition to creosote-soaked rope run along above the rows, are older remedies, but still sound practice, especially where there is a dislike of using toxic chemicals on an edible crop. Other chemicals which are available mainly to commercial growers include HCH mixtures, Chlorfenvinphos, Diazinon, Disulfoton and Phorate. **6** Carrot root aphid (see 1 above Aphis). **7** Clay burn (see Physiological disorders). **Symptoms:** carrots blotchy on the surface turning blue/black. It is not fully understood why this trouble occurs. Possibly due to lack of air in the soil. **Control:** avoid by growing in well-drained lighter soils, or improve the soil structure by working in ashes, gravel or sand. **8** Eelworm, Stem and bulb eelworm *Ditylenchus dipsaci* (see Eelworms). **Symptoms:** extreme distortion of the leaves and, in some cases, eventual death. It is noticeable that areas of plants here and there are badly affected. This eelworm can only be seen with a strong magnifying glass or microscope. Eggs may be found inside the stems or around the roots of affected plants. A wide range of different plants may be attacked. **Control:** in the case of carrots, a long rotation of suitable plants is the only real solution (see Rotation). **9** Slugs (see Slugs and snails). **10** Soft rot *Erwinia carotovora* (see Bacterial diseases). **Symptoms:** soft, mushy rotting of carrots, while growing or in store. This is more of a problem in wet areas and where land is poorly drained, and also in gardens where too much manure and fertilizers have been used. **Control:** destroy the affected carrots; practise rotation (see Rotation). **11** Splitting (see Physiological disorders). **Symptoms:** splits on carrots from crowns to tips. The causes can be varied. Excess nitrogen is said to be the main cause, but intermittent periods of dry and wet weather may also cause splitting. **Control:** little can be done, apart from making sure that soil conditions are suitable and that potash feeding has been high for late carrots. **12** Violet root rot, caused by the fungus *Helicobasidium purpureum*. **Symptoms:** blue colouration of the roots; very bad rotting. This disease is worse where conditions are colder.

Control: destroy affected carrots; practise rotation (see Rotation). **13** Virus (see Virus diseases). **Symptoms:** dwarfing of plants and great reduction in yield. **Control:** control aphids which spread the disease (see 1 above Aphis). **14** Willow and carrot aphid (see 1 above Aphis).

Caterpillars The larvae of butterflies and moths are known as caterpillars. Caterpillars vary in colour, size, shape and number of legs. Some are called loopers, because their bodies form a loop as they move. Eggs may be laid by adult butterflies and moths on plants or elsewhere. The caterpillars of most moths may feed on leaves, tunnel in leaves, bore into stems, feed on fruits or flowers, or feed on the underground parts of plants, such as roots, bulbs and tubers. They may cause serious damage to plants, and always cause fouling with their characteristic dark droppings. Butterfly caterpillars are, in general, not regarded as serious pests, with obvious exceptions, such as the caterpillars of the cabbage white butterflies, which are extremely destructive.

Control: familiarize yourself with the habits of particular moths and butterflies and use proprietary products following directions carefully. There are a number of chemicals available, such as Bioresmethrin, Carbaryl, Derris, Pyrethrum, Piperonyl butoxide and Trichlorphon.

Cauliflower or Broccoli These two names are interchangeable, but the hardier varieties are often called broccoli, and the summer varieties cauliflower. The branching varieties, with numerous small heads, are called sprouting broccoli. Calabrese is a form of sprouting broccoli. Varieties derived from *Brassica oleracea botrytis cauliflora*. Hardy and half-hardy biennials, with stems woody at the base. Height: small to medium. The heads of immature flowers are ready for cutting at various times throughout the year, according to variety and method of growing. Use: food crop, grown for heads of immature flowers. For **Pests and Diseases**, see Brassicas.

Celery Varieties derived from *Apium graveolens*. Hardy biennial. Height: medium. Leaf stalks: fleshy, green or white, according to variety and method of growing. Use: grown for the leaf stalks which, according to variety, may or may not be blanched.

Pests and Diseases. 1 Aphis *Dysaphis apiifolia* and other *Aphis* species (see Aphis). May attack young and established plants. **Symptoms:** aphids on leaves. **Control:** spray with Bioresmethrin, Derris, Dimethoate, Malathion, Oxydemeton-methyl or other insecticide, observing directions carefully. Chemicals available to commercial growers include Demeton-S-methyl. **2** Carrot fly *Psila rosae*. This pest also attacks celery (see Carrot 5 carrot fly). **Control:** dust with Calomel; spray with Malathion which gives a measure of control. Disulfoton and Phorate are recommended for commercial growers. **3** Celery fly, Celery leaf miner *Euleia* (Acidia) *heraclei*. May attack young and established plants. **Symptoms:** adult flies, about 3 mm ($\frac{1}{8}$ in) long, with nut-brown bodies and iridescent wings. They are active in May. Eggs are laid on leaves and the larvae, when they hatch out, burrow between the two leaf skins, causing blisters. There may be 2 or 3 generations in the summer. The larvae pupate either in the blisters or in the soil. Plants lose vigour and make poor growth in bad attacks. **Control:** spray with Malathion, starting in May. **4** Early blight, caused by the fungus *Cercospora apii*. **Symptoms:** circular yellow spots turning to dark brown with yellow margins, on the leaves. **Control:** spray with Benomyl. **5** Late blight or leaf spot, caused by the fungus *Septoria apiicola*. A serious disease. May attack young or established plants. **Symptoms:** brown markings on leaves and stalks which cause severe damage; black specks on the brown spots which are spores of the fungus. This disease is seed-borne, and treated seed is available, which eliminates much of the infection. **Control:** when plants are affected, spray with Benomyl, copper fungicide, Thiram or Zineb, or, as a preventive measure, spray seedlings several times with one or other of these chemicals. **6** Leaf miner (see 3 above Celery fly). **7** Leaf spot (see 5 above Late blight). **8** Phoma root rot, caused by the fungus *Phoma apiicola*. May attack at the seedling stage. **Symptoms:** collapse of seedlings as a result of attack below soil level. **Control:** use sterilized or clean soil-less composts (see Composts) clean containers and, as this disease can be seed-borne, clean seed. Provided this is done, there should be little trouble. **9** Pythium root rot, caused by *Pythium* species of fungi. May attack at the seedling stage. **Symptoms:** collapse of seedlings at soil level. **Control:** spray with copper fungicide or Thiram; always use sterilized or clean soil-less composts (see Composts), and clean containers. **10** Slugs (see Slugs and snails). **11** Soft rot *Erwinia carotovora* (see Bacterial diseases). **Symptoms:** soft rotting of the centre of the plant. This disease enters through wounds made by, for instance, celery fly attacks or by mechanical damage to plants during cultivation. Excess use of nitrogen fertilizer may make plants soft and subject to disease attacks. **Control:** control pests; avoid damage to plants; use appropriate amounts of fertilizer. **12** Virus, Mosaic virus (see Virus diseases). May attack established plants. **Symptoms:** yellow markings on the leaves; poor growth. Aphids may carry this disease to celery from carnations and irises. **Control:** control aphids (see 1 above Aphis).

Cell A microscopic, single, living unit. Cells make up organic structures and, in numbers, constitute tissue.

Cell culture (see Meristem culture)

Centipedes Centipedes are *Geophilus* and *Lithobius* species. They may be confused with millipedes (see Millipedes), but have

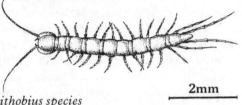

2mm

Centipede Lithobius species

flattened or thread-like, yellowish-brown bodies, one pair of legs to each body segment and antennae (feelers attached to the head). One very noticeable difference between centipedes and millipedes is that centipedes move very rapidly and, when disturbed, will try to escape. They are useful in gardens, as they feed on various soil pests such as slugs, snails and various insects, and should not be destroyed.

Certified stock (see also Approved stock). Stock with an official certificate to confirm that it is free, or relatively so, from pests and diseases and will not result in poor yield or performance (see Stock). Apples, blackcurrants, pears, plums, potatoes, raspberries and strawberries are the main plants which come under an official certificate scheme. Various grades of certification are issued according to the degree of inspection carried out by official inspectors. It may be against the law to sell plants without a certificate, as is the case in Scotland with blackcurrants, potatoes and strawberries. Sources of certified stock are available from all horticultural inspectors or advisers within the official Advisory Service in Britain.

Chafer beetles Chafer beetles include the cockchafer *Melo-*

lontha vulgaris, the garden chafer *Phyllopertha horticola* and the rose chafer *Cetonia aurata*. Both adult beetles and larvae are pests. The larvae, sometimes known as white grubs, are fat, soft, white and curved, with brown heads. The beetles vary in colour and size. They may be brown or green, and up to 25 mm (1 in) long. The beetles are most common in areas where there are heath or woodland soils which tend to be light and dry. They may be seen in large swarms in spring and early summer. The larvae feed on the roots of grass and other plants. The beetles feed by biting pieces out of leaves, young shoots, flowers or fruits of various plants.
Control: spray with Carbaryl in early summer as soon as the beetles appear; hand-pick beetles; use insecticide sprays on foliage of trees and shrubs; apply Naphthalene (if available) to the soil; trap beetles with pieces of turf.

Cheiranthus   (see Wallflower)

Chelated compounds   (see Sequestrols)

Cherry   Sweet cherry varieties derived from *Prunus avium* and sour or cooking cherry varieties derived from *Prunus cerasus*. Deciduous trees. Height: small to medium. Flowering time: spring. Flower colours: pinkish-white or pink. Fruit: cream, crimson, red or yellow. Use: food crop, grown for its fruit.

Pests and Diseases. 1 Aphis, Cherry blackfly *Myzus cerasi* (see Aphis). May attack at all stages. **Symptoms:** leaves and young shoots crowded with black aphids; leaves may be badly curled and growth checked. This may be a very damaging pest. **Control:** where practical, spray with Dimethoate, Fenitrothion or Malathion at the white bud stage. **2** Bacterial canker (see Plum 2 Bacterial canker). **Symptoms:** gum oozes from cankers. **3** Blackfly (see 1 above Aphis). **4** Blossom wilt (see Plum 3 Blossom wilt). **5** Brown rot (see Plum 4 Brown rot). **6** Bullfinch *Pyrrhula pyrrhula*. May attack at all stages. Attacks from as early as February to bud burst. **Symptoms:** bullfinches systematically strip the flower buds from each branch, causing complete loss of fruit in years following autumns in which their favourite diet of seeds is in short supply, causing them to switch to flower buds. **Control:** render the buds distasteful by frequent applications of strong sprays of either Anthraquinone or Thiram in late winter and early spring; less time-consuming but more expensive is the provision of bird-netting or cotton threads to obstruct the movement of birds through the trees. **7** Cherry bark tortrix *Enarmonia formosana*. May attack young and

69

established plants. **Symptoms:** pinkish or brownish caterpillars with black heads form tunnels in the bark, and feed on the leaves. **Control:** scrape the trunk to remove loose bark in March, or spray in the dormant season with tar oil winter wash. **8** Cherry blackfly (see 1 above Aphis). **9** Cherry fruit moth *Argyresthia curvella*. May attack established plants. **Symptoms:** eggs laid in late winter; small, green, transparent caterpillars with brown heads first seen in the open on flowers in spring, feeding on stamens and petals; later, caterpillars bore into the green fruits, doing considerable damage. **Control:** spray with tar oil winter wash in the dormant season, to destroy eggs. Commercial growers may use Azinphos-methyl, a very poisonous chemical not available to amateur gardeners. **10** Cherry leaf scab, caused by the fungus *Gnomonia erythrostoma*. May attack established plants. **Symptoms:** leaves turn brown and die. **Control:** spray with tar oil winter wash or DNOC in the dormant season. **11** Fruit tree red spider mite *Panonychus ulmi* (see Plum 7 Fruit tree red spider mite). **12** Gummosis (see Physiological disorders). May occur after hard frost. Gum oozes from areas on branches and trunk. **Control:** no control. This condition is not serious, except on trees already poor in growth. **13** Silver leaf (see Plum 15 Silver leaf). **14** Winter moth (see Plum 18 Winter moth).

Chilli   (see Pepper)

China aster   (see Aster)

Chlorosis   This happens when the normal green colouring of leaves changes to, or is patched with, pale green or yellow. Chlorosis has a number of causes, as many plant pests, diseases and disorders produce it to varying degrees.

Christmas rose   *Helleborus niger* Christmas rose and varieties, as well as *Helleborus orientalis* Lenten rose and varieties, and other *Helleborus* species and varieties. Hardy herbaceous perennials. Height: small to medium. Flowering time: winter, spring. Flower colours: green, pink, purple or white. Use: borders, woodland gardens, garden and commercial cut flower, occasionally grown as a short-term pot plant.

Pests and Diseases. 1 Leaf spot, caused by the fungus *Coniothyrium hellebori* (see Leaf spot). May attack established plants. **Symptoms:** leaves covered with large, round, black spots, causing severe damage to plant. **Control:** spray with a copper-based fungicide regularly. Note also that frost damage causes spotting of a similar sort.

Chrysanthemum   *Chrysanthemum* varieties are complex hybrids far removed from the *Chrysanthemum* species from which they are derived (see Hybrid). Chrysanthemum varieties are divided into a number of classes according to type and normal flowering time. Some Chrysanthemum varieties are suited to the modern year-round system of growing, that is, growing the full year round by extending or shortening the day length by artificial lighting or shading. There are other chrysanthemums, such as *Chrysanthemum maximum* (see Chrysanthemum maximum) and annual chrysanthemums. Greenhouse (outdoors summer/autumn) herbaceous perennials with stems woody at the base. Height: small to medium. Flower colours: cream, orange, pink, purple, red, white, yellow or bicoloured. Flowering time: throughout the year according to type and method of growing. Treated an annuals or biennials. Use: beds, borders in the open, garden and commercial outdoor cut flowers, greenhouse and commercial glasshouse cut flowers, commercial glasshouse pot plants.

Pests and Diseases. 1 Aphis, Chrysanthemum aphid *Macrosiphiniella sanborni*, Leaf-curling plum aphid *Brachycaudis helichrysi*, Mottled arum aphid *Myzus circumflexus* and other *Aphis* species (see Aphis). May attack at all stages. **Symptoms:** aphids on the plants; marking of the leaves; poor growth. **Control:** for outdoor plants use suitable sprays; for the greenhouse use smokes, aerosols and fumigation; suitable insecticides include Demeton-S-methyl commercially, gamma-HCH, Malathion and Nicotine, but HCH is not effective against the mottled arum aphid; Demeton-S-methyl has systemic properties and is effective as a soil drench for commercial use only (see Systemic), but there are reports that it has damaged a number of varieties so it must be used cautiously and on a trial basis initially; fumigation with nicotine is very effective; some species are able to transmit chrysanthemum virus diseases from diseased to healthy plants, so control of aphids should be routine practice (see Virus diseases). **2** Bishop bug *Lygus rugulipennis*. May attack young and established plants. **Symptoms:** distorted stems, leaves and flowers. **Control:** use gamma-HCH spray or smoke. **3** Caterpillars, Caterpillars of various moth species, including those of the Angleshades moth *Pholgophora meticulosa* (see Caterpillars), which are green and dark brown, with a white stripe along the back and V-shaped grey markings. Caterpillars tend to feed at night, so may not be obvious.

May attack young and established plants. **Symptoms:** holes in foliage and buds and flower petal margins bitten. **Control:** spray with Diazinon or gamma-HCH, or use HCH dusts. **4 Blotch,** caused by the fungus *Septoria chrysanthemella*. May attack established plants. **Symptoms:** rounded, dark grey or blackish spots or blisters on the leaves, especially the lower ones up to 2.5 cm (1 in) in diameter; it is most severe in conditions of excessive nitrogenous manuring and in badly ventilated greenhouses, and in these conditions plants may be defoliated. **Control:** remove and burn affected leaves and, if necessary, spray the plants with Bordeaux mixture or Zineb. **5** Chrysanthemum gall midge *Rhopalomyia chrysanthemi*. May attack established plants. **Symptoms:** stems and leaves covered with cone-shaped galls 2 mm ($\frac{1}{12}$ in) long, where the larvae feed. **Control:** spray with Diazinon or gamma-HCH at 10-day intervals. **6** Chrysanthemum leaf miner *Phytomyza syngenesiae*. May attack established plants. **Symptoms:** fine white spots on the foliage, caused by adults feeding, followed by silvery-white, snake-like markings on the leaves, caused by the larvae tunnelling inside. **Control:** spray with Diazinon or gamma-HCH. **7** Chrysanthemum stool miner *Psila nigriconnis*. May attack in stool beds. **Symptoms:** in a bad attack, no new shoots are formed, but even a slight attack may impair the vigour of plants and the production of cuttings; larvae feed on roots and tunnel in the rootstocks and the bases of the stems. **Control:** spray stool beds with gamma-HCH twice in late summer which will effectively control this pest. **8** Earwig *Forficula auricularia*. May attack established plants. **Symptoms:** ragged marks on buds and flowers where earwigs have fed. **Control:** spray with gamma-HCH; use traps made of boards, tiles, sacks, cans or flower pots. **9** Eelworm (see 14 below Leaf and bud eelworm). **10** Flower scorch, caused by an *Intersonilia* species of fungus. May attack established plants. This disease has recently come into prominence and in some years may cause very severe loss of blooms. **Symptoms:** first seen in the outer florets as small rusty brown spots, or as larger roughly oval translucent spots; affected florets wither and turn brown and soon the whole flower is withered and brown; grey mould (see 13 below Grey mould) may mask the early symptoms. This disease may occur in the greenhouse when humidities are high. In damp autumns it may severely damage outdoor chrysanthemums. **Control:** the use of various fungicides has given variable results and is still at the investigational stage; to help prevent the spread of disease, raise the temperature in the greenhouse. **11** Glasshouse red spider mite *Tetranychus urticae*. May attack young and established plants. The mite has a small, pear-shaped body, yellow-green or red in colour. All stages of the pest are visible. **Symptoms:** foliage hard and parchment-like, with yellow mottling on the upper surface turning completely yellow; webs are produced on plants and spoil flowers. **Control:** use one of the many acaricides available, among them Dimethoate and Demeton-S-methyl commercially; for young plants use Tetradifon; for mature plants commercially, use any one of these with Dicofol added, alternatively, use Azobenzene or Tetradifon smokes. **12** Glasshouse whitefly *Trialeurodes vaporariorium* (see Whitefly). May attack young and established plants. **Symptoms:** small adults with white wings feed in large numbers on young plants and on the growing points of established plants, rising in clouds if the foliage is disturbed; dark areas develop on the leaves where they feed, and leaves are covered with honeydew and sooty moulds (see Honeydew); the very small larvae look like scales attached to the plants. **Control:** spray with Diazinon or Malathion; biological control is possible (see Biological control). **13** Grey mould, caused by the fungus *Botrytis cinerea* (see Botrytis). May attack established plants. **Symptoms:** water-soaked spots on the petals and rotting of the flowers; stems and leaves may rot; cuttings are also affected. **Control:** after flowering is finished, remove and destroy all old plants, including roots and any other waste; wash down the greenhouse with disinfectant or detergent; fumigate with Formalin or sulphur (not in metal houses); sterilize containers and canes; sterilize soil before planting again (see Soil sterilization); after planting, spray Captan on stem bases and adjacent ground and repeat 3 weeks later, or spray Maneb on stem bases but *not* on adjacent ground; a stem base spray of Benomyl after planting ($\frac{1}{2}$ litre per plant, 1 pint per plant) with subsequent whole plant sprays have been found to be good; avoid any checks to plant growth and conditions of high humidity; spray with Benomyl after deleafing; inspect plants after coming into flower and remove any showing signs of infection; Tecnazene smokes may be used and Dichlofluanid sprays are recommended. **14** Leaf and bud eelworm *Aphelenchoides ritzemabosi* (see Eelworms). May attack young and established plants. **Symptoms:** leaves show yellow-brown blotches, which enlarge and turn black, until leaves finally wither;

buds may also turn black. **Control:** remove all withered leaves; destroy plants after flowering; remove all infested soil; use healthy plants only for propagation and give hot water treatment before taking cuttings (see Hot water treatment). **15** Leaf miner (see 6 above Chrysanthemum leaf miner). **16** Leafy gall *Corynebacterium fascians* (see Bacterial diseases). May attack young and established plants. **Symptoms:** plants produce masses of short, thickened or distorted shoots at the bases of the stems and distorted buds. As this disease is very infectious, care should be taken to prevent infected sap being transmitted from infected to healthy plants on hands or knives. **Control:** destroy infected stools or plants; sterilize containers, implements and soil which have been in contact with infected plants; avoid planting chrysanthemums in ground which has recently carried a diseased crop; do not take cuttings from diseased plants. **17** Mineral deficiencies (see Mineral deficiencies). Chrysanthemums are sensitive to various mineral deficiencies. **Symptoms:** include yellowing of the leaves, scorching of the leaves and general poor growth. To establish what the deficiency is and how to correct it, consult the local horticultural adviser. **18** Petal blight, caused by the fungus *Itersonilia perplexans*. May attack established plants. **Symptoms:** pin-point-sized reddish spots on the petals coalesce to form water-soaked blotches, the tips of the outer florets being affected at first, then the whole bloom. **Control:** use warmth and ventilation to reduce humidity; if necessary, spray commercially with Zineb, preferably freshly prepared from Nabam and zinc sulphate; spray at coloured bud stage, and repeat at weekly intervals. **19** Powdery mildew caused by the fungus *Oidium chrysanthemi* (see Mildew). May attack established plants. **Symptoms:** powdery white growth on the surface of the leaves. **Control:** spray with Benomyl, colloidal sulphur, Dinocap or commercially with Quinomethionate at 10 to 14-day intervals; drench soil with Dimethirimol. **20** Ray blight, caused by the fungus *Mycosphaerella ligulicola*. May attack at all stages. **Symptoms:** dark marks on cuttings; dark blotches on stems and leaves of older plants; dark brown areas in the centres of flowers which spread outwards, often affecting one side of the flower only. **Control:** spray regularly with Captan, Maneb or commercially with Mancozeb; after clearing out old plants, sterilize border soil, if possible using steam, as the value of chemical sterilization (see Soil sterilization) is still uncertain. **21** Root rots. (a) Root rot, caused by a

Phoma species of fungus. May attack established plants. **Symptoms:** stunted plants with yellowing of lower leaves; stems may crack at soil level; pink flecks on roots, which rot. **Control:** commercially drench the soil with Nabam at 7 to 10-day intervals, in place of the normal watering; use Captan or Zineb in the same way; after clearing out old plants, and before planting again, sterilize border soil thoroughly or commercially drench soil with Nabam (see Soil sterilization). (b) Root rot, caused by the fungi *Fusarium* species, *Phytophthora* species, *Rhizoctonia solani*, *Thielaviopsis basicola*. **Symptoms:** rotting of roots of older plants. **Control** (see (a) above). **22** Rust, caused by the fungus *Puccinia chrysanthemi* (see Rust). May attack established plants. **Symptoms:** yellowish-green spots on the upper surfaces of the leaves, rusty brown pustules beneath. **Control:** spray regularly with Thiram or Zineb when the disease first appears; if the disease occurs commonly, spray as a routine. **23** Stem rot, caused by the fungus *Sclerotinia sclerotiorum*. **Symptoms:** light brown markings on the stems which become covered with whitish fungal growth in which small hard black bodies (sclerotia) may be embedded; they may also be present inside the stems. **Control:** use sterilized soil. **24** Tarnished plant bug (see 2 above Bishop bug). **25** Virus diseases caused by various viruses (see Virus diseases). They include for example: (a) Aspermy, Flower distortion or Tomato aspermy virus. **Symptoms:** flowers smaller than normal and slightly or severely distorted; breaking of the flower colour (breaking is the term used when *part* of a flower changes colour). **Control:** control aphids, which spread the disease, with insecticidal sprays (see Aphis); take cuttings from healthy plants only: (b) Chrysanthemum stunt or Stunt virus. **Symptoms:** whole plant greatly reduced in size; in some varieties, leaf spotting or flecking occurs; flowers may develop earlier than normal, and may be of poor quality and colour. **Control:** destroy affected plants; take cuttings from healthy plants only: (c) Other virus diseases. **Symptoms:** various degrees of mottling or death of the leaves, discolouration and/or malformation of the flowers. **Control:** examine plants regularly and destroy plants showing virus symptoms on leaves or flowers; routine insecticidal spraying should be done to control aphids and thrips which can spread virus diseases. **26** Whitefly (see 12 above Glasshouse whitefly). **27** Wilt, caused by the fungi *Verticillium albo-atrum*, *Verticillium dahliae*. **Symptoms:** lower leaves turn yellow, then brown, and wilt; symptoms spread up the plant; woody stem bases of plants may be brown, and outer layer of stems may separate easily from centres. **Control:** if disease appears in isolated patches, remove affected plants and apply Benomyl drench to soil around neighbouring plants; sterilize border soil before planting again; take cuttings from healthy plants only.

Chrysanthemum maximum and varieties. Hardy herbaceous perennial. Height: medium to large. Flowering time: summer. Flower colours: cream, white. Use: herbaceous borders, garden and commercial cut flower.

Pests and Diseases. 1 Aphis (see Aphis). May attack established plants. **Symptoms:** aphids and honeydew with sooty moulds make plants unsightly (see Honeydew). **Control:** seldom a serious trouble, and spraying is unnecessary. **2** Blotch, caused by the fungus *Septoria leucanthemi*. May attack established plants. **Symptoms:** blotching on the leaves. Not really a serious disease. **Control:** if troublesome, pick off infected leaves and spray with a copper fungicide. **3** Earwig *Forficula auricularia*. May attack established plants. **Symptoms:** distorted flowers. **Control:** trap with inverted flower pots on canes; spray regularly with HCH if necessary.

Cineraria Varieties derived from *Senecio cruentus*. For convenience, they are put into groups according to flower and plant size. Greenhouse herbaceous perennial with stems woody at the base. Height: small to medium. Flowering time: varies according to time of sowing, from winter to spring and summer. Flower colours: blue, cream, pink, purple, red, white or parti-coloured. Use: greenhouse and indoor pot plants. Usually treated as an annual or biennial.

Pests and Diseases. 1 Aphis, various species (see Aphis). May attack young and established plants. **Symptoms:** aphids, honeydew and sooty moulds on plants, making them unsightly and checking growth (see Honeydew). **Control:** spray with Malathion. **2** Chrysanthemum leaf miner *Phytomyza syngenesiae*. May attack young and established plants. **Symptoms:** pale, winding tunnels in the lowest leaves. **Control:** provided the damage is quickly detected, careful pinching of the widest part of each tunnel is often sufficient control (see Chrysanthemum 6 Chrysanthemum leaf miner). **3** Glasshouse red spider mite *Tetranychus urticae* (see Mites). May attack young and established plants. **Symptoms:**

discoloured leaves; honeydew and sooty moulds making the plants unsightly and checking growth (see Honeydew). **Control** (see Mites). **4** Glasshouse whitefly *Trialeurodes vaporariorum* (see Whitefly). May attack at all stages, but usually worst on established plants in late summer. **Symptoms:** whiteflies, honeydew and sooty moulds on leaves (see Honeydew). **Control** (see Whitefly). *Note* that in the case of cineraria, whiteflies breed so rapidly that destruction of the plants is often the only recourse. **5** Grey mould, caused by the fungus *Botrytis cinerea* (see Botrytis). May attack established plants. **Symptoms:** fluffy grey fungal growth on leaves and poor growth of plants. **Control:** remove and destroy diseased leaves; avoid overcrowding plants and maintain suitable temperatures and ventilation. **6** Leaf miner (see 2 above Chrysanthemum leaf miner). **7** Powdery mildew caused by an *Oidium* species of fungus (see Mildew). May attack young and established plants. **Symptoms:** powdery fungal growth on the leaves. **Control:** commercially Dimethirimol formulations may be effective; avoid overwatering plants. **8** Red spider (see 3 above Glasshouse red spider mite). **9** Virus, Tomato spotted wilt virus (see Virus diseases). **Symptoms:** yellow spotting on leaves followed by vein discolouration and finally death of plants. **Control:** remove and destroy any plant showing symptoms immediately; spray to control insects which may spread this disease. **10** Whitefly (see 4 above Glasshouse whitefly). **11** Wilt, caused by a *Phytophthora* species of fungus. May attack young and established plants. **Symptoms:** rotting of roots and bases of stems, and wilting of plants. **Control:** discard badly affected plants; always use sterilized composts and clean containers.

Cissus *Cissus antarctica* Kangaroo vine or Russian vine. Greenhouse (outdoors in the very mildest areas only) evergreen shrub. Height: climbing, medium to tall, according to use as a border or pot plant. Foliage: decorative. Growth: decorative. Use: borders or pots in the greenhouse, indoor pot plants.

Pests and Diseases. 1 Aphis (see Aphis). May attack established plants. **Symptoms:** aphids, honeydew and sooty moulds on plants spoiling appearance of the foliage (see Honeydew). **Control:** spray with Malathion. **2** Broad mite *Hemitarsonemus latus* (see Mites, Begonia 3 Broad mite). **3** Leaf scorch (see Physiological disorders). May attack established plants. **Symptoms:** browning and withering of leaves; poor general state of plants. This is almost entirely due to poor growing conditions, and especially to excess light. **Control:** improve growing conditions; avoid excess light.

Citrus fruits Including the lemon *Citrus limonia*, the sweet orange *Citrus sinensis*, and their varieties, and the ornamental *Citrus mitis*. Evergreen trees (greenhouse in Britain). Height: small. Flowering time: throughout the year. Flowers: white and fragrant. Use: grown for their decorative flowers, fruit and foliage.

Pests and Diseases. 1 Palm thrips *Parthenothrips dracaenae*. May attack established plants. **Symptoms:** adults with brown bodies and yellow legs on the foliage. **Control:** spray with Diazinon, Malathion or Nicotine. **2** Soft scale *Coccus hesperidum* (see Scale insects, Soft scale).

Clarkia *Clarkia elegans* and varieties. Hardy annual. Height: medium. Flowering time: summer. Flower colours: pink, purple, red, white. Use: borders outside, pots in the cold greenhouse.

Pests and Diseases. 1 Downy mildew, caused by the fungus *Peronospora arthuri* (see Mildew). **Symptoms:** lower leaves coated with a fine, dry mould; death of upper leaves. Whole plants may be affected. **Control:** spray with a copper fungicide, Maneb or Mancozeb commercially, Thiram or Zineb; before sowing, dip seed in Thiram.

Coleus *Coleus blumei* and varieties. Half-hardy and greenhouse perennial. Flowers: insignificant. Leaves: brilliantly patterned and coloured. Use: greenhouse and indoor pot plant, summer bedding, window boxes and outdoor containers. Usually an annual.

Pests and Diseases. 1 Citrus mealybug (see Mealybugs, Camellia 4 Mealybug). **2** Slugs and snails (see Slugs and snails). **3** Woodlice (see Woodlice). May be especially troublesome with young plants.

Columbine *Aquilegia* species and hybrids. Hardy herbaceous perennials. Height: small to medium. Flowering time: summer. Flower colours: blue, cream, pink, purple, white, yellow or particoloured. Foliage: fern-like and graceful. Use: herbaceous borders, rock gardens, woodland gardens.

Pests and Diseases. 1 Leaf spot, caused by several species of fungi. May attack established plants. **Symptoms:** white or grey-brown spots on leaves. **Control:** spray with a copper fungicide or Captan. **2** Rust, caused by the fungus *Puccinia agrostidis* (see Rust). May attack established plants. **Symptoms:** orange-coloured pustules on undersides of leaves; distorted leaves. **Control:** spray with Thiram.

Composts   The term for pest and disease free mixtures of materials used to supply food, water and support to a plant in any container used in plant culture. (Note that the term is also used for waste organic material rotted down to provide soil fertilizer.) There are three different categories of composts. 1 Soil-based, or Seed or Potting Composts. 2 Soil-less Composts or growing media. 3 Aggregates and Substrates or inert materials. 1 Soil-based composts usually contain soil, other organic matter, and added nutrients. The best examples of soil-based composts are the John Innes Seed and Potting Composts, formulated by the John Innes Institute. These can be bought already made up but they should not be kept for long. They build up ammonia and are toxic to plant growth (see Osmosis, Soluble salts). However, it is perfectly safe to use John Innes composts bought from a reliable source, provided they are not kept more than a few months. If you are mixing your own composts, follow the instructions below. For sowing seeds, make a Seed Compost, containing little plant food with: 2 parts by bulk sterilized loam, 1 part by bulk peat, 1 part by bulk gritty sand. To each bushel (this can be most easily measured in a box $56 \times 25 \times 25$ cm or $22 \times 10 \times 10$ in), add 20 g ($\frac{3}{4}$ oz) ground limestone or chalk and 40 g ($1\frac{1}{2}$ oz) superphosphate of lime. For young plants, make No. 1 mixture Potting Compost with: 7 parts by bulk sterilized loam, 3 parts by bulk peat, 2 parts by bulk gritty sand. To each bushel, add 20 g ($\frac{3}{4}$ oz) ground limestone and 110 g ($\frac{1}{4}$ lb) John Innes base. (To make John Innes base, mix 2 parts by weight hoof and horn meal, 2 parts by weight superphosphates and 1 part by weight sulphate of potash.) For larger plants, double the amount of lime and John Innes base. This will make No. 2 mixture Potting Compost. For mature and greedy plants, treble the amount of lime and John Innes base. This will give No. 3 mixture Potting Compost. 2 Soil-less composts contain organic matter and nutrients such as peat or peat plus sand, vermiculite, perlite or other relatively inert aggregate and nutrients. These composts are used in growing bags (growing modules or bolsters). The advantage of soil-less composts is that sterilization is not necessary as they should be free from pests and diseases. Their disadvantage is that they are unable to retain nutrients in the way that soil-based composts can. All the plant nutrients added to them are taken up by the plants relatively quickly or are washed out by watering. A wide variety of ready mixed soil-less composts are available, but if you are mixing your own composts, follow the instructions below. For a simple soil-less compost, mix the following: 1 part by bulk peat (ideally spongy brown peat moss, not black decomposed peat, although a small proportion of the latter is useful) 1 part by bulk gritty sand. To each bushel, add 170–230 g (6–8 oz) ground limestone and a base fertilizer, such as Vitax Q4, Osmocote, Plantasan or Enmag, in the quantity or proportion recommended by the manufacturers. All of the above base fertilizers contain trace elements (see Trace elements), but it is advisable if in doubt, to add these, for example Frit 253A, which is a readily available source of chelated trace elements. Alternatively, supplement the fertilizer with a liquid plant food containing trace elements. Plants in soil-less composts must be fed regularly with liquid fertilizers, preferably containing trace elements. 3 Aggregates and Substrates. These are inert materials, that is, they do not contain plant food, and are used in hydroponics. Generally, they do not need to be sterilized (see Soil sterilization). Examples are ashes, lignite (brown coal), mineral wools, perlite, plastics (polystyrene), vermiculite and sand. As they do not contain plant nutrients, all feeding must be applied in the form of a complete liquid food.

Convallaria   (see Lily of the valley)

Coral spot   (*Nectria cinnabarina*). A very common fungus which affects many different woody plants (shrubs and trees especially). Generally saphrophytic, attacking dead tissue, but can also affect living tissue in some cases.

Symptoms: masses of red-coloured cushions or clusters of spores on dead wood. It is by these spores that the disease is spread.

Control: remove dead tissue or wood back to live growth and protect wound with paint; avoid bad pruning snags which die and serve as a convenient point of entry for spores; do not allow infected material to lie about.

Cordyline   Closely related to *Dracaena*, with which it is often confused. *Cordyline australis* and varieties, *Cordyline indivisa* and *Cordyline terminalis* and varieties. Greenhouse (some outdoors in mildest areas only), evergreen trees and shrubs. Height: in pots in the greenhouse, small shrubs; in the open, medium shrubs to small trees. Use: greenhouse pot plants, outdoors in containers or borders in very mild areas only. For **Pests and Diseases**, see Dracaena.

Crataegus   (see Hawthorn)

Cress (see also Mustard and cress). Varieties derived from *Lepidium sativum*. Greenhouse annual. Height: as grown, dwarf. Use: food crop, cut at the seedling stage for garnishing or salads.

Pests and Diseases. 1 Damping off, caused by *Phytophthora* and *Pythium* species of fungi (see Damping off). May attack at the seedling stage. **Symptoms:** turning brown of seedling stems at soil level; collapse of seedlings. **Control:** water with copper fungicide or dust with Quintozene; always use sterilized soil in composts and clean containers.

Crop rotation (see Rotation)

Cross A plant raised by crossing 2 distinct cultivars or varieties (see Cultivar, Variety) is called a cross. The term may sometimes be used when 2 species are crossed (see Cross-pollination).

Cross pollination Flowers may be pollinated by their own pollen (self-pollination) or they may be pollinated by the pollen of a flower of the same variety or species, or of another variety or species (cross-pollination). Cross pollination is the rule in most plants, though some are self-pollinated, or are able to self-pollinate if necessary. The deliberate and controlled pollination of flowers with pollen from another variety or species is called crossing, and produces hybrids (see Hybrid) for new cultivars or varieties (see Cultivar, Variety).

Cuckoo-spit bugs The young stages of a species of froghopper *Philaenus spumarius* are called cuckoo-spit bugs or cuckoo-spit

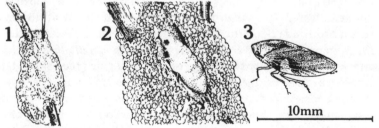

1 White froth   2 Young stage feeding on shoot   3 Adult

insects. They are yellow, and live in a white froth which they make. They are common pests of various garden plants, such as lavender, rose and a number of different herbaceous plants. Cuckoo-spit bugs attack plants in early summer, feeding mainly on young shoots and flower stalks.

Control: first, hose the plants vigorously with water, to clear away the froth, then spray with gamma-HCH, Malathion or a systemic insecticide and repeat the spray as necessary.

Cucumber *Cucumis sativus* varieties, including the ridge cucumber. Half-hardy and greenhouse herbaceous perennial. Height: climbing. Flowering time: throughout the year, according to method of growing. Flower colour: yellow. Fruit: green. Stems: rough with tendrils. Use: food crop grown for its fruit, borders or containers in greenhouses or commercial glasshouses, frames or other structures, ridge cucumbers in the open in good summer areas.

Pests and Diseases. 1 Aphis *Aphis gossypii* (see Aphis). May attack at all stages. **Symptoms:** yellowing of foliage; distorted fruitlets; yellowish-green to black aphids with long feelers on the plants. **Control:** commercially spray with Demeton-S-methyl, or Dimethoate; established plants are also tolerant of sprays of Malathion and smokes of Nicotine. **2** Bees, Honey bee *Apis mellifera*. Various wild bee species. May affect established plants. **Symptoms:** swollen, bitter-tasting fruit. Cross-pollination, caused by the bees, results in bitter, deformed fruit. **Control:** use screened ventilators or regularly remove all male flowers. Varieties without male flowers are now available. **3** Black root rot, caused by the fungus *Phomopsis sclerotioides* (see Root rot). May attack young and established plants. **Symptoms:** wilting and death of plants; rotted stem bases and roots, with black spots on the outer tissues, which strip off easily; blackening of older rotted roots. **Control:** discard affected plants; sterilize border soil (see Soil sterilization) or remove soil and re-soil the border; grow the plants in straw bales. **4** Black stem rot, caused by the fungus *Macosphaerella melonis*. May attack established plants. **Symptoms:** small pale green spots, later turning brown, on the leaves; edges of the leaves have a water-soaked appearance; dark marks on the stems where leaves or side-shoots have been removed; soft, greyish-green rot at the tips of the fruits. **Control:** cut out dead or diseased growth; spray with Thiram or Zineb at the first sign of disease, then at 14-day intervals, or dust with Thiram; reduce humidity in the greenhouse; when plants are finished, thoroughly cleanse the greenhouse and destroy all debris. **5** French fly *Tyrophagus neiswanderi*. May attack young and established plants. This pest is a mite (see Mites) not a fly, and white, globular mites with large bristles are visible on the plants. Severe infestations cause blindness of the shoots. **Symptoms:**

minute holes in young foliage which enlarge as the foliage grows. Mites are brought into the greenhouse on straw of horse manure. In high temperatures they become active and migrate to young plants, where they start to feed. **Control:** spray with Diazinon or Dicofol; avoid using straw or manure which may be infested with mites. **6** Fruit disorders (see Physiological disorders). May occur on established plants. **Symptoms:** distorted or split fruit; white or pale brown corky scars on the fruit; marked failure of fruit to develop. **Control:** avoid checks to plant growth, such as cold draughts or sudden drops in temperature, when cucumbers are forming; ensure suitable soil conditions for good root development. **7** Fungus gnats *Bradysia* species. May attack seedlings and young plants. The larvae are 6 mm ($\frac{1}{4}$ in) long, translucent, with black heads. **Symptoms:** larvae feed on the root hairs of the seedlings, slowing growth or causing collapse; they may also feed on the leaves, making small holes. **Control:** drench beds or containers with Malathion; give extra water; prior to sowing seed, treat the soil with Diazinon. **8** General plant disorders (see Physiological disorders). May occur on young and established plants. **Symptoms:** blind shoots; fasciation (flattened and fused shoots); scorching of leaves either marginal or between the veins; temporary wilting of the plants. Causes of these troubles are not fully known. **Control:** grow plants as well as possible and avoid damage to tips of shoots, which can cause fasciation. **9** Glasshouse red spider mite *Tetranychus urticae* (see Mites). May attack at all stages and, if not controlled, will kill all young growth. **Symptoms and control** (see Tomato 18 Glasshouse red spider mite) but *do not use* Tetradifon on cucumber seedlings. **10** Glasshouse whitefly *Trialeurodes vaporariorum* (see Whitefly). May attack at all stages. **Symptoms and control** (see Tomato 20 Glasshouse whitefly) but Diazinon and Malathion sprays may be used *only on older* cucumber plants. **11** Grey mould, caused by the fungus *Botrytis cinerea* (see Botrytis). May attack established plants. **Symptoms:** water-soaked marks at stem joints and at the scars where leaves and side-shoots have been removed. **Control:** remove infected parts; reduce humidity; spray with Benomyl or Thiophanate-methyl, or dust with Captan. **12** Gummosis, caused by the fungus *Cladosporium cucumerinum*. May attack established plants. **Symptoms:** on young fruits, grey sunken spots which exude a sticky fluid; splitting of fruits, exposing inner tissues; green, fungal masses on the affected tissue. **Control**: destroy affected fruit; reduce humidity; avoid low temperatures in the greenhouse; spray with Captan, Thiram or Zineb at 10-day intervals; grow resistant varieties (see Resistant varieties). **13** Millipedes, Glasshouse millipede *Oxidus gracilis* (see Millipedes). May attack at all stages. Millipedes are brown, with flattened bodies, about 8 mm ($\frac{1}{3}$ in) long and with numerous legs. Unlike centipedes, they move slowly and, when disturbed, curl up tightly. **Symptoms**: millipedes gnaw stems just above soil level, destroying seedlings; on older plants, roots are attacked and the damage done may allow entry of diseases. **Control**: drench soil with gamma-HCH; always use sterilized compost for seedlings; thoroughly clean the greenhouse after old plants have been removed. **14** Powdery mildew, caused by the fungus *Erysiphe cichoracearum* (see Mildew). May attack established plants. **Symptoms**: white, powdery mould on the leaves. **Control**: spray with Dinocap, colloidal sulphur, Quinomethionate commercially, Thionate-methyl or Benomyl at 10 to 14-day intervals; commercially, drench the soil with Dimethirimol. **15** Red spider (see 9 above Glasshouse red spider mite). **16** Rootknot eelworm *Melodogyne hapla* (see Eelworms). May attack established plants. **Symptoms** (see Tomato 33 Rootknot eelworm). **Control**: as for tomatoes, but no resistant rootstocks are available; culturing on straw bales or in peat reduces the incidence of eelworm attack. **17** Seedling disorders (see Physiological disorders). **Symptoms**: uneven germination; collapse of seedlings and distortion of seed leaves. **Control**: use reliable seed; avoid using too much fertilizer in seed composts; maintain suitable temperatures for germination, 27°C (80°F); handle seedlings carefully to avoid damage; water carefully. **18** Springtails *Collembola* species. May attack seedlings. The minute white or colourless insects may be seen in large numbers if some soil is floated in a bucket of water. **Symptoms**: skeletonization of leaves; destruction of seedlings. Serious for seedlings only. **Control**: drench soil with gamma-HCH. **19** Stem and root rot, caused by fungi of *Phytophthora* species, *Pythium* species and *Rhizoctonia* species (see Stem rot, Root rot). May attack seedlings and young plants. **Symptoms**: markings of light brown (*Rhizoctonia*) or dark brown (*Phytophthora* or *Pythium*) at the bases of stems of seedlings or young plants; rotting of shoots. **Control**: always sterilize border and container soil; if *Rhizoctonia* occurs commercially, apply Quintozene dust to the surface of the soil and rake in; for *Phytophthora* and *Pythium*, remove diseased plants and apply copper fungicide to the bases of unaffected plants. **20** Virus diseases (see Virus diseases). (a) Green mottle, Cucumber green mottle virus. **Symptoms:** inconspicuous light and dark green mottling on the young shoot leaves; no symptoms on older leaves or fruits; marked loss of yield can occur. **Control:** destroy infected young plants by watering pots with disinfectant such as Formalin or cresylic acid 2% solution; in the border, carefully remove and destroy affected plants; the disease is easily transmitted by handling, so always wash hands carefully; it is also transmitted by tools, containers and other items so wash these in trisodium phosphate 5% solution; disease is seed-borne, but seed can be freed of infection by heating it to 70°C (158°F) for 3 days. (b) Mosaic, Cucumber mosaic virus. **Symptoms:** yellow and light green pattern on the leaves, with star and ring markings; distortion of leaves and fruit. Symptoms vary according to the strain of virus occurring (see Strains). **Control:** destroy young infected plants; as the disease is easily transmitted by handling, always wash hands, and do not handle healthy plants after infected ones; spray with insecticide to control aphids which can carry the disease from infected to healthy plants. **21** *Whitefly* (see 10 above Glasshouse whitefly).

Cultivar This term indicates CULTIvated VARiety. It is now used for a distinct cultivated variety of a plant with individual features or qualities. Confusion exists between the words 'variety' and 'cultivar', and while, in essence, the two are synonymous, the purist will maintain that 'cultivar' should only be used where a plant has been produced by contrived or deliberate breeding, whereas 'variety' should be used where a plant arises naturally as a variant from the natural species. Broadly speaking, it is not really important which term is used, and in recent years, there has been less use of the word 'cultivar'. Throughout this book the term 'variety' is used for cultivated varieties. Note that cultivated varieties include hybrids (see Hybrid) and sports. The word 'sport' applies when part of a plant changes, usually in one character only, for example, a pink-flowered rose may produce a branch with flowers which are similar in every way, except that they are white. Sports or bud variations of this sort may be propagated, usually vegetatively, and the resulting plant given a variety name.

Currants (see Blackcurrant, Red and white currants).

Cutworms Cutworms, also called surface caterpillars, are the larvae of different kinds of moth including the garden dust moth *Euxoca nigricans*, the heart and dart moth *Agrotis exclamationis*, the turnip moth *Agrotis segetum*, the yellow underwing moth *Noctua pronuba* and other species. The caterpillars have varying numbers of legs according to species, and vary in colour from grey-white to greenish-grey. They live in the surface of the soil and come out at night to feed, cutting plants off at ground level and causing wilting and death. Severe attacks can be very damaging, especially to seedlings and young plants. **Control:** apply Bromophos or poison baits to the soil (see Poison baits); hoe regularly as it exposes cutworms to birds during the day in the months when they are active. All weeds, on cultivated ground and elsewhere, should be kept down, as eggs are laid and larvae later feed on many common weeds as well as on cultivated plants.

Cyclamen *Cyclamen persicum* Glasshouse cyclamen and varieties, *Cyclamen neapolitanum* and other *Cyclamen* species. Hardy and greenhouse herbaceous perennials with tuberous roots. Height: dwarf to small. Flowering time: throughout the year, according to sort and method of growing. Flower colours: pink, purple, red or white. Flowers: may be fragrant. Leaves: may be patterned and zoned. Use: greenhouse pot plants, short term indoor pot plants, commercial glasshouse pot plants, hardy sorts borders, rock gardens, woodland gardens. Note that the following pests and diseases mainly affect glasshouse cyclamens.

Pests and Diseases. 1 Aphis, Mottled arum aphis *Aulacorthum circumflexum*. May attack young and established plants. **Symptoms:** plants covered with bright green aphids and honeydew (see Honeydew). **Control:** spray with Demeton-S-methyl commercially, or with Malathion. **2** Black root rot, caused by the fungus *Thielaviopsis basicola*. May attack established plants. **Symptoms:** softening and yellowing of the leaves; failure of the plants to flower; black and rotted roots. **Control:** destroy affected plants; use sterilized compost and clean containers; water at intervals with Captan or Benomyl. **3** Cyclamen mite *Steneotarsonemus pallidus*. May attack young and established plants. Mites are very difficult to see. **Symptoms:** outer leaves curl upwards; flower buds wither and rot. **Control:** spray with Dicofol, repeating spray several times at intervals. **4** Grey mould, caused by the fungus *Botrytis cinerea* (see Botrytis). May attack established plants. **Symptoms:** grey mould on the leaves; rotting of leaf stalks; spotted flowers; if severe, will kill the leaves and check growth. **Control:** spray with Benomyl; improve ventilation. **5** Vine weevil *Otiorhynchus sulcatus* (see Begonia 13 Vine weevil). **6** Weevil (see 5 above Vine weevil).

D

Daffodil Name given to *Narcissus* species and varieties which have trumpet flowers. Narcissus is the name used for species and varieties which have flowers with large or small cups. Daffodil and narcissus varieties are complex hybrids far removed from the *Narcissus* species from which they were derived (see Hybrid). A few *Narcissus* species are grown, chiefly in rock gardens. Narcissus species and varieties are divided into classes, according to flower form, species and species from which varieties have been derived. Mostly hardy, herbaceous perennials, with bulbs. Height: small to medium. Flowering time: varies with species and variety, and with method of growing, from winter to early summer. Flower colours: cream, orange, salmon-pink, white, yellow and combinations of these colours. Use: beds, borders, naturalized in grass, rock gardens, garden and commercial cut flower, commercial glasshouse or indoor pot plant, window boxes and outdoor containers.
Pests and Diseases. 1 *Aphis*, various aphis species (see Aphis). **Symptoms:** distortion of leaves; restricted growth. As well as doing direct damage, aphids may spread certain virus diseases (see Virus diseases). **Control:** spray with Malathion, Derris or Dimethoate. **2** Basal rot, caused by the fungus *Fusarium oxysporum*. **Symptoms:** bulbs initially soft with rot at base; spreading of chocolate-brown rot upwards, with white or pinkish material between the scales, which is part of the fungal growth. Infected bulbs either rot away, or produce a few stunted, yellowish leaves. **Control:** discard any soft bulbs; handle bulbs carefully and store in a cool place. Commercial growers add formalin (0.5%) in the hot

water treatment against eelworm (see Hot water treatment, Eelworm). **3** Blindness (see Physiological disorders). **Symptoms:** no flowers emerge; stunted or poorly developed shoots. This may be due to incorrect temperatures during storage of the bulbs adversely affecting the flower buds within the bulbs. Or it may be due to starting to force commercial glasshouse, greenhouse or indoor bulbs too soon, before the bulbs are adequately developed. **Control:** follow correct routines in storage and forcing. **4** Bulb mite (see Bulb mite). **5** Bulb scale mite (see Bulb scale mite). **6** Dry set (see Physiological disorders). **Symptoms:** papery and shrivelled flowers. This may be due to incorrect temperatures during storage of the bulbs adversely affecting the flower buds within the bulbs. Or it may be due to forcing commercial glasshouse, greenhouse or indoor bulbs at too high a temperature. **Control:** follow correct routines in storage and forcing. **7** Eelworm (see **12** below Stem and bulb eelworm). **8** Fire, caused by the fungus *Sclerotinia polyblastis*. **Symptoms:** small brownish spots on the flowers; longish spots on leaves, which may be killed under very damp conditions. **Control:** cut flowers, even if they are not wanted, before they fade, to prevent spread of infection; rake off and remove old leaves when they have died down; spray with Bordeaux mixture during the growing season, though this may do some damage to foliage. **9** Narcissus flies: large, *Merodon equestris*, small, *Eumerus strigatus* and *Eumerus tuberculatus*. **Symptoms:** failure of bulbs to produce shoots or production of numerous thin and grassy leaves; soft bulbs and rotted basal plates. A longitudinal cut through an affected bulb will show a feeding cavity with either one single plump and yellow maggot, or several small white maggots. **Control:** destroy affected bulbs; always buy good bulbs from a reliable source which will have had hot water treatment which destroys narcissus fly larvae (see Hot water treatment). **10** Slugs (see Slugs and snails). **11** Smoulder, caused by the fungus *Botrytis narcissicola*. **Symptoms:** spotting of leaves and flowers; in severe cases, rotting of emerging shoots; small flat black sclerotia (resting bodies of the fungus) under the outer scales causing bulbs to rot. Cool, humid conditions encourage this disease. Infection is less severe in warm, dry weather. **Control:** remove and destroy badly infected plants; spray remaining plants with Zineb to prevent the disease spreading; always inspect bulbs before planting and discard diseased ones; if disease has been present in the greenhouse, sterilize border soil and soil for pot bulbs. **12** Stem and bulb eelworm *Ditylenchus dipsaci* (see Eelworms). A serious pest. **Symptoms:** lopsided leaves bearing flaky scars on growing plants in spring; twisted, puffy and horribly distorted stems; green petals on flowers; soft and silvery bulbs; whitish woolly substance exuding from neck of bulbs. **Control:** destroy affected bulbs; buy good quality bulbs and plant on fresh ground. Commercial growers use hot water treatment, soaking bulbs for 3 hours at 44°C (112°F) (see Hot water treatment), but this is not suitable for bulbs intended for forcing, or bulbs are dipped in a solution of Thionazin (0.23%). **13** Stripe or yellow stripe (see Virus diseases). **Symptoms:** longitudinal pale to yellow stripes on the leaves; rough surface on the leaves in some varieties after infection. **Control:** remove plants showing symptoms during the early growth stages.

Dahlia *Dahlia* varieties are complex hybrids far removed from the *Dahlia* species from which they are derived (see Hybrid). Dahlia varieties are divided into a number of classes according to flower form and size. Half-hardy and greenhouse herbaceous perennials with tuberous roots. Height: small to large. Flowering time: summer. Flower colours: cream, orange, pink, purple, red, white, yellow or parti-coloured. Use: borders, summer bedding, garden and commercial outdoor cut flowers, window boxes and outdoor containers, short-term greenhouse and indoor pot plants.

Pests and Diseases. 1 Aphis, Black bean aphid or Blackfly *Aphis fabae* (see Aphis). May attack young and established plants. **Symptoms:** unsightly aphids on plants causing distortion. Aphids may spread virus diseases (see Virus diseases) as well as damaging plants, so must be controlled. **Control:** spray with Demeton-S-methyl commercially, Derris, HCH, Malathion, Nicotine or Oxydemeton-methyl. **2** Capsid bug, Common green capsid bug *Lygus pabulinus*. May attack early in the growing period but damage may not become apparent until flowering time. **Symptoms:** adult bugs, bright green in colour and 6 mm ($\frac{1}{4}$ in) long, move rapidly and jerkily over the plant, puncturing leaves and flower buds; their feeding marks develop as holes or brown corky scars and flowers become distorted. **Control:** where damage has occurred in previous years, spray with Carbaryl, Dimethoate or gamma-HCH in June or early July and do not wait until damage is seen; remove woody plant refuse, in which the eggs overwinter. **3** Caterpillars, various species. May attack young and established plants.

Symptoms: leaves eaten. **Control:** spray with Derris. **4** Earwig *Forficula auricularia.* May attack young and established plants. **Symptoms:** leaves and flowers distorted by earwigs feeding. **Control:** spray regularly with HCH; trap with pots inverted on stakes. **5** Glasshouse red spider mite *Tetranychus urticae* (see Chrysanthemum 11 Glasshouse red spider mite). **6** Grey mould, caused by the fungus *Botrytis cinerea* (see Botrytis). May attack established plants. **Symptoms:** grey mould on leaves, flowers and stems, followed by rotting; tubers in store may also be attacked and may rot. This disease may be very troublesome in wet seasons. **Control:** spray with Benomyl; grow plants in suitable soil and position, and avoid overcrowding; dahlia tubers may often be attacked in the ground in cold, wet weather, when conditions are unfavourable, so lift tubers out carefully, to avoid any mechanical damage, and keep in a cool, dry, frost-proof store. **7** Leaf gall *Corynebacterium fascians* (see Bacterial diseases). May attack at all stages. **Symptoms:** gall-like swellings and dense leafy shoots near ground level. This disease may attack a number of different kinds of plants. **Control:** remove and destroy affected plants or cuttings; do not propagate from diseased plants. **8** Leaf spot, caused by the fungus *Entyloma dahliae.* May attack established plants. **Symptoms:** round spots with yellow-brown margins on the leaves; streaks on the stems and leaf stalks; if a leaf is held up to the light, dark spores will be visible. **Control:** destroy badly affected plants; spray with Bordeaux mixture. **9** Red spider mite (see Chrysanthemum 11 Glasshouse red spider mite). **10** Rosy rustic moth (see Antirrhinum 7 Rosy rustic moth). **11** Thrips, Onion thrips *Thrips tabaci.* May attack established plants. **Symptoms:** distortion of leaves and flowers caused by thrips feeding. **Control:** spray with Derris, HCH or Malathion. **12** Verticillium wilt, caused by the fungus *Verticillium albo-atrum* (see Wilt). May attack established plants. **Symptoms:** apparently healthy plants suddenly begin to wilt; dark markings at the bases of the stems; if stems are cut, centres will be dark. **Control:** practise rotation as this disease will remain in the soil (see Rotation); ground may be sterilized with Basamid, but rotation is the best control; sterilized composts and containers should be used during propagation; in severe cases destroy all plants, and start again with healthy plants on clean ground. **13** Virus diseases (see Virus diseases). Several serious virus diseases affect Dahlias.

Symptoms: mottling, patterning or spotting of the foliage; general poor growth and poor flowering; stunting of the plants. (a) Cucumber mosaic virus. **Symptoms:** spotting and distinct light green areas on the leaves. (b) Dahlia mosaic. **Symptoms:** banding in a chevron pattern on the leaves and stunting of the plants. It is useless to keep virus infected plants. **Control:** lift and destroy all infected plants; start again with new healthy plants from a reputable source; spray regularly to control aphids and thrips which can spread virus diseases (see 1 above Aphis, 11 above Thrips).

Damping off Damping off is a general term applied to various plant diseases or disorders, but mainly to fungal attacks on seedlings

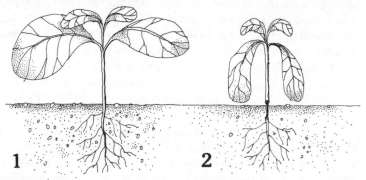

1 Healthy plant   2 Attack at base of stem

or cuttings. Attacks on slightly older plants are usually described as foot rot or stem rot (see Foot rot, Stem rot) and on established plants, as wire stem. Damping off may occur in the open, though it is more common in the greenhouse, because of the humid atmosphere. Symptoms are collapse of stems at soil level, and death of seedlings or cuttings. In older plants, stems turn thin and hard at soil level and may rot, so crippling the plants. Damping off is caused by various fungi, including *Phytophthora* species, *Pythium* species, *Rhizoctonia solani* and *Thielaviopsis basicola* which enter the stems at soil level. In the greenhouse, damping off is encouraged by unfavourable growing conditions, such as unsuitable seed or cutting composts (especially soil based types with excess ammonia gas), overcrowding in seed and cutting containers, overwatering and excess humidity. In the open, damping off is encouraged by unfavourable growing conditions such as damp or acid soils, poor preparation of the seedbed and overcrowding of seedlings.

Control: in the greenhouse, clean containers and sterilize composts; use soil-less media (see Soil-less media); maintain good growing conditions; water seedlings with Cheshunt compound, Quintozene or Thiram or commercially with Aaterra or Milcol; in the open, sterilize the seedbed with Basamid or commercially, with a liquid form of Metham sodium before sowing seed, allowing the recommended time before sowing to elapse (see Soil sterilization); practise rotation (see Rotation).

Damson   (see Plum)

Daphne   *Daphne mezereum* and other *Daphne* species. Hardy deciduous and evergreen shrubs. Height: small. Flowering time: spring, summer. Flower colours: cream, lilac-blue, purple-red, white, yellow or yellowish-green. Flowers usually very fragrant. Berries: decorative, blue-black, red, white or yellow. Use: planted singly or in groups in borders, rock gardens.

Pests and Diseases. 1 Virus, several virus diseases, including Cucumber mosaic virus (see Virus diseases). Serious diseases. May attack established plants. **Symptoms:** pale yellow mottling of foliage, with dead, brownish spots, rings and flecks; stunted plants. Daphnes are largely infected with virus diseases, which are difficult to control. **Control:** buy plants which appear to be healthy, and try as far as possible to minimize the severe check which always affects daphnes at planting time; spray plants to control aphids, which can spread virus diseases (see Aphis).

Deficiency diseases and disorders   (see Physiological disorders, Mineral deficiencies)

Delphinium (Larkspur)   *Delphinium* species and varieties. Commonly, the tall perennials are known as Delphinium and the smaller annuals as Larkspur. Hardy annual, biennial or perennial herbaceous plants. Height: small to tall. Flowering time: summer to autumn. Flower colours: blue, cream, pink, purple, red, white, yellowish or parti-coloured. Use: borders, herbaceous borders, rock gardens, commercial cut flowers.

Pests and Diseases. 1 Black blotch, caused by the fungus *Pseudomonas delphinii*. May attack established plants. **Symptoms:** jet black spots on leaves. This disease is spread from spores in the ground. **Control:** spray with Bordeaux mixture early in the season. **2** Cutworms (see Cutworms). May attack young and established plants. **Symptoms:** new shoots eaten off at ground level. **Control:**

use poison baits (see Poison baits). **3** Delphinium moth *Polychrisia maneta*. **Symptoms:** caterpillars feeding on buds, leaves, flowers and seeds from April to June; a web is visible, binding stems and leaves together. **Control:** cut out parts affected; spray with HCH or other insecticide. **4** Mildew, caused by the fungus *Erysiphe polygoni* (see Mildew). May attack established plants. **Symptoms:** white mould on leaves, which makes them wither prematurely, and on buds, which affects flowering. **Control:** cut back badly affected plants to the ground; spray early next season with liquid copper fungicide, Dinocap, Quinomethionate or sulphur as limesulphur. **5** Slugs (see Slugs and snails). A particularly troublesome pest. May attack young and established plants. **Symptoms:** young shoots eaten in spring. **Control:** put down poison baits (see Poison baits); before shoots come through ground in early spring, some clean, old, well-weathered ashes over delphinium crowns will help to discourage slug attack.

Dracaena   (see also Cordyline). *Dracaena fragrans* and varieties, *Dracaena marginata* and varieties, and other *Dracaena* species and varieties. Greenhouse, evergreen, tree or shrub-like plants. Height: small to medium as pot plants. Flowers: small, greenish-yellow. Foliage: very decorative in colour and form. Use: greenhouse pot plants requiring high temperatures, indoor pot plants for short spells when temperatures are suitable.

Pests and Diseases. 1 Bay scale *Dynaspidiotus britannicus* (see Scale insects). May attack young and established plants. **Symptoms:** light brown convex scales on the stems and leaves. **Control:** spray with Diazinon at 14-day intervals. **2** Leaf spot, caused by the fungus *Macrophoma draconis*. May attack established plants. **Symptoms:** dark spots on the leaves. **Control** (see Leaf spot). **3** Mealybug, Root mealybugs *Rhizoecus* species (see Aloe 1 Mealybug). **4** Palm thrips *Parthenothrips dracaenae*. May attack young and established plants. **Symptoms:** discoloured foliage; very small, yellow and brown adult thrips on the plants. **Control:** spray with Malathion or use gamma-HCH dusts or smokes. **5** Root mealybug (see 3 above Mealybug). **6** Scale insects (see 1 above Bay scale). **7** Thrips (see 4 above Palm thrips).

Drenches   Plants affected by soil-living fungi or pests can be given a drench of a suitable chemical (see Appendix p. 245–54). Systemic pesticides may be applied in the same way to control pests and diseases on the leaves.

Drought Prolonged period without any rainfall which increases the stress on plants endeavouring to obtain water and nutrients from the soil (or the material the plants are growing in). The effect of drought varies according to soil type, level of natural water table, rainfall in the area, and many other factors. Plants vary in their ability to withstand stress and some, such as cacti, are specially adapted to do so. Large trees with deep searching roots are able to seek water from lower depths. If no water at all is supplied, the plant cells will cease to function, and gradual or quick death of the plant (depending on species and ability to withstand lack of water) takes place.

Dusts Several pesticides are available in 'puffer-packs' for use as dusting powders. These are particularly useful in droughts, when a spray may cause scorch (phytotoxicity), but the dust deposit temporarily disfigures the plant. Dusting powders should not be used in windy conditions because of the possibility of drift.

E

Earwigs The common earwig *Forficula auricularia* should be looked upon as a beneficial insect, as it feeds on aphids and other small pests, but it occasionally lapses and causes plant damage. The insect is of moderate size and has a pair of pincers at the hind end. As a pest, the earwig is best known for its habit of climbing into the flower heads of dahlias and chrysanthemums, where it chews the tips of the florets, and in orchards, it cuts neat circular holes in mature fruit, which may lead to rotting.

Control: the earwig cannot fly so it can be prevented from climbing fruit trees by the use of grease banding material or barriers of insecticide (impregnated sacking); with flower crops, the insects may be shaken from the plants; bad infestations can be controlled by application of gamma-HCH, as a spray or as a dust, to the soil and to the lower parts of the plant, at weekly intervals; it is also important to remove breeding places of the earwig.

Eelworms These are extremely small, usually microscopic, though some may be seen with a hand lens. They are worm-like in form, though not closely related to worms. There are a number of different species, many of which are very serious plant pests.
Symptoms: raised spots and bands on leaves; distortion of leaves, buds and flowers; softening and rotting of bulbs; abnormal growth of tubers and small nodules on roots. Laboratory examination of roots is required to confirm the presence of eelworm. Different species live in or on various parts of plants, and are also able to persist in the soil for a number of years. With potato cyst eelworm, dead females, known as cysts, are visible on potato roots and tubers. These pinhead-sized, creamy or dark brown, rounded cysts each enclose several hundred eggs containing living larvae, which can remain viable in the soil for many years. All that is needed for the minute larvae to hatch out and invade root tissue is for the root exudations to be stimulated. **Control:** if plants in the garden are known to have been attacked by eelworm, destroy them at once; do not use propagating material taken from infested plants; practise long rotation (see Rotation). Commercial growers use various treatments for different kinds of eelworm, such as bulb dips, hot water treatment (see Hot water treatment), but these methods require knowledge, skill and often specialized equipment if they are to be safe and effective. (See also Soil sterilization.)

Egg plant (see Aubergine)

Elm *Ulmus procera* English elm and varieties, *Ulmus glabra* Wych elm and varieties, and other *Ulmus* species and varieties. Hardy deciduous trees. Height: small to large. Varied habits of growth, including weeping. Foliage: may turn golden-yellow in autumn, variegated in some sorts. Use: planted singly or in groups or avenues in large gardens, parks or open land.
Pests and Diseases. 1 Coral spot, caused by the fungus *Nectria cinnabarina*. May attack established plants. **Symptoms:** red spots on branches which have died back. This disease may spread to live wood, but this is not generally the case. **Control:** check why branches are dying back, and improve conditions, if possible. **2** Dieback, caused by various species of fungi, may also be physiological (see Physiological disorders). May attack established plants. **Symptoms:** branches die back and drop. **Control:** if practical, cut branches back, and fill wounds with cement; establish if the general growing conditions are suitable and check drainage.

3 Dutch elm disease, caused by the fungus *Ceratocystis ulmi*. May attack young and established plants. A serious disease. **Symptoms:** badly scorched leaves, with branches dying from the tips; whole branches usually die, or the foliage appears to be thin. This is a very serious disease, which has received considerable publicity. Report this disease, if suspected, to the Forestry Commission in Britain and have it confirmed. It usually affects trees between 10 and 40 years old. The disease is spread by elm bark beetles (see 4 below Elm bark beetles). **Control:** older trees are generally removed immediately, but there is a treatment for younger trees, based on the chemical Carbendazim for which results are mixed. **4** Elm bark beetles, *Scolytus* species. May attack young and established plants. **Symptoms:** female beetles bore holes in tree bark and form burrows in which to lay their eggs, doing considerable damage to the trees and also spreading Dutch elm disease. **Control:** the only protective measure is to fell and burn affected trees. **5** Verticillium wilt, caused by the fungus *Verticillium albo-atrum*. May attack established plants. **Symptoms:** wilting down one side of the trees; damage appears similar to that caused by Dutch elm disease. The infection usually comes from the soil, from a previous crop such as potatoes. **Control:** after confirming the disease, little can be done beyond removing the tree before it becomes dangerous.

Erica (see Heath)

Euonymus *Euonymus* species and varieties. Hardy deciduous and evergreen shrubs and small trees. Height: prostrate shrubs to small trees, climbing. Flowers: insignificant. Fruit: often decorative in colour and form. Fruit colours: orange, pink or red. Foliage: variegated may be tinted in purples, reds or yellows in spring or autumn. Use planted singly or in groups in borders, hedges or climbers against walls.

Pests and Diseases. 1 Black bean aphid *Aphis fabae* (see Aphis). May attack established plants. **Symptoms:** black aphids and eggs on leaves and twigs. **Control:** spray with Malathion. **2** Spindle ermine caterpillar *Yponomeuta cognatella*. May attack established plants. **Symptoms:** caterpillars in cobwebby masses feeding on leaves. Euonymus is likely to be attacked in early summer. **Control:** spray two or three times with Derris or Trichlorphon.

Euphorbia (see Poinsettia)

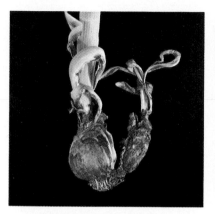

Daffodil **Stem and bulb
eelworm** *Ditylenchus dipsaci*,
distorted stem and bulb (see
page 89)

Elm **Elm bark beetles**
Scolytus species, spreader of
Dutch elm disease, female
beetles boring into bark (see
page 96)

Fuchsia **Rust** caused by the
fungus *Uredo fuchsiae*,
undersides of leaves showing
reddish pustules (see page
102)

Greenhouse **Mealybugs**
covered with their
characteristic white waxy
substance

Greenhouse **Whitefly** attack
many plants in the greenhouse

Greenhouse **Vine weevil**
larvae feed inside buds,
flowers, fruit, shoots, stems
and roots

Hyacinth **Yellow disease**
Xanthomonas hyacinthi, yellow
streaks on leaves (see page
116)

Hydrangea **Capsid** damage
caused by *Lygocoris pabulinus*,
weakened and distorted leaves
(see page 117)

F

F1 Hybrid   (see Hybrid)

Fagus   (see Beech)

Family   In plant classification, a family is a large group of one or more genera that have general similarities (see Genus).

Ficus   *Ficus elastica* Rubber plant varieties, other *Ficus* species and varieties and *Ficus carica* Common fig (see Fig, common). Greenhouse, evergreen trees and shrubs. Height: trees, in nature small to large; shrubs, small to medium, climbing and trailing. Leaves: decorative, varying in colour, shape, size and variegation according to sort. Use: greenhouse border and pot plants, indoor pot plants, commercial glasshouse pot plants. Note that the common fig, which has a decorative foliage and fruit, may be grown outside in mild areas.

Pests and Diseases. 1 Leaf discolouration (see Physiological disorders). May occur on young and established plants. **Symptoms:** browning of leaf margins. May be due to overwatering especially in winter time. **Control:** water with care. **2** Leaf dropping (see Physiological disorders). May occur on established plants. **Symptoms:** lower leaves drop off. May be due to underwatering. **Control:** water correctly at all times. **3** Scale insects, Hemispherical scale *Saissetia fici*, Soft scale *Coccus hesperidum* (see Scale insects). **Symptoms:** leaves and stems covered with whitish or brownish scales and honeydew (see Honeydew); poor growth, with plants wilting and possibly dying. **Control:** spray with Malathion.

Fig, common   *Ficus carica* and varieties. Greenhouse (in the open in mild areas only), evergreen tree. Height: small. Flowering time: over a period, spring to summer. The so-called fruits which are green or purple, are in fact the flowers. Foliage: decorative (see also Ficus). Use: food crop grown for the 'fruit' but also an ornamental plant, greenhouse borders, containers, outdoor borders in mild areas only, best against walls.

Pests and Diseases. 1 Canker, caused by the fungus *Phomopsis cinerescens*. May attack established plants. **Symptoms:** canker wounds on the branches, with bark roughened at sites of infection; branches may die if the cankers are extensive. **Control:** cut off and burn diseased branches, ensuring that no snags are left where infection might lodge; paint cut areas with a suitable protective paint, for example white lead paint. **2** Dieback and fruit rot or Grey mould, caused by the fungus *Botrytis cinerea* (see Botrytis). May attack established plants. **Symptoms:** wilting and death of young shoots; rotting of fruits, which may remain mummified on the trees. **Control** (see Tomato 23 Grey mould). **3** Eelworm, Fig cyst eelworm *Heterodera fici* (see Eelworms). May attack established plants. **Symptoms:** lemon-shaped cysts (see Eelworm), varying from white to brown in colour, on the roots. **Control:** lift and destroy affected trees; replace with healthy young plants from a reliable source; sterilize soil before replanting. **4** Grey mould (see 2 above Dieback and fruit rot). **5** Scale, Soft Scale *Coccus hesperidum* (see Scale insects). May attack established plants. **Symptoms:** yellow-brown, long-oval scales on the midribs on the undersides of the leaves. **Control:** spray with Malathion.

Fig, ornamental   (see Ficus)

Firs   (see Silver firs)

Flea beetles or Turnip fly   A group of beetles, mainly *Phyllotreta* species. They are small, active jumping beetles, varying

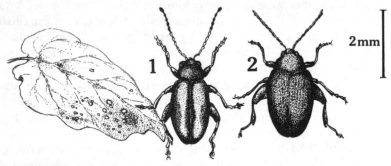

1 Phyllotreta nemorum   2 Psylliodes affines

in size and colour according to species. They are usually blue-black, often with two yellow stripes on their backs. Adult beetles overwinter in rubbish heaps, under tree bark and in other such places and come out in spring, flying to various plants, where they feed and lay eggs. Seedlings, especially of brassicas, suffer badly and may be destroyed, so that re-sowing is necessary (see Brassicas).

The leaves of older plants are holed and netted, but they can usually survive attacks.
Control: spray with Carbaryl, Derris or Nicotine, or commercially with Tetrachlorvinphos, altering frequently the chemicals used to avoid build-up of resistance in beetles.

Foot rot Foot rot is a general term used to describe discolouration and rotting of the plant stem at or near ground level. It is usually applied to fungal diseases of herbaceous plants, especially seedlings, cuttings and young plants. Many different fungal diseases may cause foot rot in a wide range of plants. The term foot rot is often used together with the term root rot, and a plant may be said to have foot and root rot, when both the roots and the base of the stem are affected. Foot rot may also be called stem rot, and plants may be said to have stem and root rot, when both stems and roots are affected.
Control: use sterilized composts or soil-less media, clean containers and clean benches in the greenhouse; take great care over hygiene at the propagating stage; maintain good growing conditions; commercially, water plants with fungicides such as Milcol. It is important to establish what is causing the foot rot, so that the most appropriate control may be used.

Forget-me-not *Myosotis caespitosa* Bedding forget-me-not and varieties, *Myosotis scorpioides* Forget-me-not and other Myosotis species and varieties. Hardy, chiefly biennials, perennials. Height: small. Flowering time: spring. Flower colours: blue, pink or white. Use: spring bedding, waterside and woodland gardens, containers in the open and window boxes, rock gardens.
Pests and Diseases. 1 Downy mildew, caused by the fungus *Peronospora myosotidis* (see Mildew). **2** Powdery mildew, caused by the fungus *Erysiphe horridula* (see Mildew).

Forsythia *Forsythia suspensa* and varieties, and other *Forsythia* species and varieties. Hardy deciduous shrubs. Height: medium. Flowering time: spring. Flower colours: golden-yellow or lemon-yellow. Use: planted singly or in groups, shrub borders, trained to walls, or fences, trailing down banks.
Pests and Diseases. 1 Dieback, caused by the fungus *Sclerotinia sclerotiorum*. May attack established plants. **Symptoms:** branches die back; small black fruiting bodies of the fungus on the wood. Botrytis often follows as a secondary trouble (see Botrytis). **Control:** cut out and burn affected wood. **2** Galls and fasciation.

May attack established plants. Various causes, such as bird droppings, fungal diseases or mechanical damage acting as irritants. **Symptoms:** galls and distorted, flattened stems. Not a serious trouble. **Control:** cut out and burn all affected branches if necessary. **3** Leaf spot, caused by the fungus *Phyllosticta forsythiae* and other species of fungi. May attack established plants. **Symptoms:** spotting of leaves. **Control:** no real damage caused, but, if thought necessary, spray with a liquid copper fungicide.

Fraxinus (see Ash)

Freesia *Freesia refracta* hybrids (see Hybrid). Greenhouse (in the open in summer only) herbaceous perennials, with corms. Height: small. Flowering time: varies according to method of growing from spring to winter. Flower colours: copper, cream, lilac, orange, pink, red, white or yellow. Flowers may be fragrant. Use: greenhouse pot plants, in borders outside in summer, chiefly commercial glasshouse cut flowers. *Note* that Freesia pests and diseases apply also to Gladiolus.

Pests and Diseases. 1 Aphis, various Aphis species (see Aphis). May attack at all stages. **Symptoms:** aphids on the plants; discoloured leaves; weakened growth. **Control:** use systemic chemicals as soil drenches rather than as sprays to avoid scorching flowers and destroying their perfume (see Systemic). **2** Bulb mite *Rhizoglyphus callae* (see Bulb mite). Attacks mainly noticed on dormant corms. **Symptoms:** small glistening white mites found in soft areas beneath the papery scales, excavating holes in the flesh of corms. These mites only attack corms and bulbs previously injured by other means. **Control:** place corms in sealed tins with the moth repellant paradichlorobenzene to eliminate the mite, but this will not necessarily solve the main problem. **3** Core rot, caused by the fungus *Botrytis gladiorum*. May attack corms. **Symptoms:** centres of corms may rot and corms may be completely destroyed. **Control:** destroy infected corms; commercially protect corms by dusting with Quintozene; use sterilized compost; sterilize border soil (see Soil sterilization). **4** Corm rot (see 6 below Fusarium yellows and corm rot). **5** Dry rot, caused by the fungus *Sclerotinia gladioli*. May attack corms. **Symptoms:** small black marks on the flesh of corms which may extend and destroy them; leaves turn yellow and rot at soil level, leaving a stringy dry rot; minute black fruiting bodies of the fungus may be seen on the rotted leaves and also on the corms. **Control:** destroy all infected corms; use sterilized compost; sterilize border soil (see Soil sterilization). **6** Fusarium yellows and corm rot, caused by the fungus *Fusarium oxysporum*. May attack corms. **Symptoms:** yellowing of the leaves; zoned brown marks on the surface of the corms; reddish-brown marks inside the corms. **Control:** destroy infected corms; dip sound corms in Benomyl suspension before storing them in a dry, well-ventilated store to protect them; use sterilized compost; sterilize border soil (see Soil sterilization); with freesias grown from seed, use a Benomyl soil drench, starting when the seedlings have produced 4 leaves, and continue at 4 to 6-week intervals to reduce disease attacks. **7** Scab and neck rot *Bacterium marginatum* (see Bacterial diseases). May attack established plants. **Symptoms:** red-brown specks on the leaves which develop into elongated spots; pale yellow spots on the corms which turn black. **Control:** dip corms in solution of Calomel. **8** Thrips, Gladiolus thrips *Thrips simplex*. May attack established plants. **Symptoms:** sap oozing out where thrips have been feeding, causing a silvering and stickiness on the leaves; small dark brown adults may be visible. **Control:** spray with Malathion at 14-day intervals.

French marigold   (see Marigold)
Froghoppers   (see Cuckoo-spit bugs)
Frost and cold   Various species of plants respond very differently to frost and cold, according to whether they are hardy, half-hardy or tender. The first symptom is invariably wilting in the case of those soft or herbaceous plants deprived of water (the water being frozen and unobtainable). Wilting is followed by dieback as the cells are ruptured. This sort of damage is common in tender bedding plants, planted before the last frost in spring, or damaged during their hardening off or acclimatization period. Large woody plants have considerable sap reserve in the trunk and may not react so quickly. Splitting of bark due to cell rupture is another symptom of extreme cold. Frequently and especially with conifers, the first sign of damage is browning, burning or scorching, followed by gradual dieback. Often the symptoms of cold occur some time after the period of extreme cold. Sudden thaw can often be more damaging than extreme cold, or for that matter intermittent cold and thaw. The effect of a sudden thaw is to blacken the leaves or flowers, and often botrytis (see Botrytis) or soft rot quickly moves in to attack the damaged tissue. Many species, even those wild and hardy are reduced to ground level by cold, or a certain part is damaged. It is

advisable to cut back to a live growing bud to prevent the diseases, which attack dead tissue, from spreading to healthy tissue.

Fuchsia *Fuchsia magellanica* and varieties, and other *Fuchsia* species and varieties. Greenhouse, half-hardy and some hardy in mild areas, trees and shrubs. Height: trees, small; shrubs, small to medium. Flowering time: summer, autumn or winter, according to sort and method of growing. Flower colours: blue, lilac, pink, purple, red, white or parti-coloured. Use: hardy species, in the open, planted singly or in groups, borders, semi-wild gardens, wild gardens; half-hardy, summer bedding, outdoor containers, hanging baskets, window boxes; greenhouse, indoor pot plants, greenhouse pot plants, in hanging baskets, planted in borders and trained up pillars or rafters, hanging flowers being very decorative. **Pests and Diseases. 1** Grey mould, caused by the fungus *Botrytis cinerea* (see Botrytis). May attack established plants. **Symptoms:** under very damp conditions, flowers may rot. **Control** (see Botrytis). **2** Rust, caused by the fungus *Uredo fuchsiae*. **Symptoms:** reddish pustules on undersides of leaves; pale spots on upper surfaces. **Control:** spray with a liquid copper fungicide, Maneb, Thiram or Zineb. Dichlofluanid may also be useful. **3** Whitefly *Trialeurodes vaporariorum* (see Whitefly). May attack established plants. **Symptoms:** leaves curling and covered with honeydew (see Honeydew). **Control:** spray repeatedly with Bioresmethrin, Diazinon or Malathion.

Fungal diseases Poor or abnormal growth, or death, in plants may be due to a number of different factors, such as attack by pests, infection by bacterial, fungal or virus diseases, or by physiological disorders (see Bacterial diseases, Physiological disorders, Virus diseases). Fungal diseases are the most common diseases in plants. Fungi form one of the large groups into which the plant kingdom is divided, a group which contains both useful and harmful members. Mushrooms, toadstools, various moulds and many plant disease organisms are included in this group. Fungi, like bacteria, contain no chlorophyll, the green colouring matter of plants which enables them in the presence of light, along with other factors to make food, and fungi have therefore to depend for food on living or dead organic matter. Fungi reproduce by minute spores and build up in a tube-like system of growth, which may penetrate plant tissue, or may exist on the surface of plants. This system of growth is called mycelium, and may be seen in certain cases as a mat or felt on the surface of plants. When spores are produced, they may give a fluffy appearance, black, grey, whitish or some other colour according to the species of fungus. The fungi causing plant diseases are as a rule parasites living in or on plant tissue and deriving food from it. Some fungi are however saprophytes, living on plant tissue already dead. Some are able to grow on both dead and living tissue. Fungi may attack plants in the open or in the greenhouse. They are able to overwinter in various ways, as spores, mycelium, sclerotia (small resting bodies able to withstand adverse conditions and to start growth again once these alter) or rhizomorphs, which are fungal strands formed into string-like bodies, black, brown or whitish in colour. Different species have different and specific resting habits which help in identifying them. Spread of fungal diseases is by different methods according to type.

Control: use clean plants, free from disease; provide hygienic and good growing conditions, and practise sterilization where necessary (see Soil sterilization); use preventative chemicals which are toxic to diseases but not to the plants; use therapeutic chemicals to kill the diseases but not the host plant (see Host); use resistant varieties of plants (see Resistant varieties).

G

Galanthus (see Snowdrop)

Gardenia *Gardenia jasminoides* and varieties. Greenhouse ever-green shrub. Height: medium. Flowering time: summer to autumn. Flower colour: white. Flowers very fragrant. Use: greenhouse borders or pots. Once much more used as a cut flower than now.

Pests and Diseases. 1 Chlorosis (see Physiological disorders). May occur on established plants. **Symptoms:** leaves turn yellow. **Control:** keep the soil on the acid side (less than pH 7) and the temperature above 20°C (68°F); if the chlorosis is thought to be due to iron deficiency (see Mineral deficiencies) caused by too high a pH, apply chelated iron as Sequestrine (see Sequestrols). **2** False red spider mites *Brevipalpus* species. May attack young and established plants. **Symptoms:** greyish areas on the leaves where mites are feeding. **Control:** spray every 14 days with Dicofol or Malathion. **3** Glasshouse red spider mite *Tetranychus urticae.* May attack young and established plants. **Symptoms:** discoloured areas on the leaves where mites are feeding. **Control:** spray with Dicofol or Malathion every 14 days. **4** Grey mould, caused by the fungus *Botrytis cinerea* (see Botrytis). **5** Mealybug, Citrus mealybug *Planococcus citri* (see Mealybugs). May attack young and established plants. **Symptoms:** white spiny bugs, honeydew and sooty moulds on the leaves (see Honeydew). **Control:** spray with Diazinon or gamma-HCH. Do not spray open blooms or coloured buds. **6** Red spider (see 3 above Glasshouse red spider mite). **7** Stem canker, caused by the fungus *Phomopsis gardeniae.* May attack established plants. **Symptoms:** brown areas on stems near soil level, at first sunken but later swelling and giving a rough cracked surface; branches also affected; plant may be stunted and may even die. **Control:** destroy infected plants; spray with liquid copper to help prevent infection; when cutting blooms, do not leave snags which may become infected; take cuttings from healthy plants only and use a sterilized rooting medium.

Genus In plant classifications, a genus (pl. genera) is a group of species (see Species) with similarities in flower, fruit and seed, which indicate that they might originally, in the distant past, have been derived from one plant. A genus may contain one species, a few species, or a very large number of species, which may look very different, but are in fact closely related.

Geranium The name geranium is often incorrectly applied to members of the Pelargonium family (see Pelargonium). Various geranium species and varieties, some called by the common name Crane's-bill. Hardy herbaceous perennials. Height: small to medium. Flowering time: summer. Flower colours: blue, pink, purple, red-purple or white, often veined with a contrasting colour. Use: herbaceous borders, semi-woodland gardens and rock gardens.

Pests and Diseases. None in particular.

Gladiolus Gladiolus varieties are complex hybrids derived from various *Gladiolus* species (see Hybrid). They are, for convenience, put into classes, according to flower size, time of flowering and species from which they have been derived. Half-hardy, a few species hardy, herbaceous perennials, with corms. Height: small to medium. Flowering time: spring, summer or autumn according to sort and to method of growing. Flower colours: cream, lilac, orange, pink, purple, red, white, yellow or parti-coloured. Use: summer bedding and borders in the open, greenhouse border plants and pot plants, commercial cut flowers in the open and in the glasshouse. For **Pests and Diseases**, see Freesia.

Glasshouse whitefly (see Whitefly)

Gloxinia Gloxinia varieties are complex hybrids derived from various *Sinningia* species (see Hybrid). Greenhouse herbaceous perennials with tuberous roots. Height: small. Flowering time: long flowering period throughout the year, according to the method of growing. Flower colours: cream, blue, lilac, pink, purple, red, white or parti-coloured. Use: greenhouse pot plant, short-term indoor pot plant.

Pests and Diseases. 1 Aphis, including mottled arum aphis *Aulacorthum circumflexum* (see Aphis). May attack at all stages. **Symptoms:** leaves speckled and distorted. **Control:** spray with suitable insecticides. **2** Cyclamen mite *Tarsonemus pallidus*. May attack at all stages when plants are growing actively. **Symptoms:** death of leaves and greening of flowers. **Control:** gloxinias are sensitive to the chemicals suitable for cyclamen mite control, so it is best to destroy all infested plants. **3** Foot rot and corm rot, caused by the fungi *Phytophthora cryptogea* and *Phytophthora parasitica*. May attack established plants. **Symptoms:** rotting at base of plant which spreads into corm and up into leaves and flower stalks. **Control:** destroy all affected plants. **4** Mites (see 2 above Cyclamen mite).

Gooseberry Varieties derived from *Ribis grossularia*. Hardy deciduous shrub. Height: small. Flowering time: spring. Fruit: green, red, yellow or white, smooth or hairy. Use: food crop, grown for its fruit.

Pests and Diseases. 1 American gooseberry mildew caused by the fungus *Sphaerotheca mors-uvae* (see Mildew). May attack established plants. **Symptoms:** white fungal coating on leaves, shoots and fruits, which later turns brown and felt-like, checks growth and spoils fruit. **Control:** spray as the first flowers open with Benomyl, Dinocap or Thiophanate-methyl and repeat at 10-day intervals; if troublesome after flowering, apply Dinocap every 4 to 5 days. **2 Aphis,** Gooseberry aphid *Aphis grossulariae.* May attack at all stages. **Symptoms:** aphids and honeydew on the leaves (see Honeydew); leaves mottled and distorted. **Control:** use tar oil winter wash in the dormant season; if gooseberry bryobia mite is present, see 9 below Gooseberry bryobia mite; use DNOC at bud breaking stage; if winter wash is not effective, spray with Derris. **3 Borer.** May be Currant clearwing (see Blackcurrant 10 Currant clearwing) or Currant shoot borer (see Blackcurrant 12 Currant shoot borer). **4 Botrytis,** caused by the fungus *Botrytis cinerea* (see Botrytis). May attack established plants. **Symptoms:** dark areas on leaves, shoots and fruits, covered with fluffy white mould. This may be a very persistent and damaging disease in wet seasons. **Control** (see 1 above American gooseberry mildew). **5 Bryobia mite** (see 9 below Gooseberry Bryobia mite). **6 Capsid,** Common green capsid *Lygocoris pabulinus* (see Blackcurrant 7 Capsid). **7 Caterpillars** (see 10 below Gooseberry sawfly, 13 below Magpie moth). **8 Gooseberry aphid** (see 2 above Aphis). **9 Gooseberry bryobia mite** *Bryobia ribis.* May attack young and established plants. **Symptoms:** silvering of leaves, which may be severe in hot summers. **Control:** if mite damage is seen, spray with Derris, Dimethoate or Malathion; use a DNOC winter wash. **10 Gooseberry sawfly** *Nematus ribesii.* May attack young and established plants. Eggs are laid in April/May in batches of 20 to 30 on the undersides of the leaves. Caterpillars, green, spotted with black, and with black heads, hatch out in May. There may be several generations in the year. **Symptoms:** caterpillars feed first round the margins of the leaves; sometimes they begin feeding in the centre of bushes and eat outwards; in bad attacks the bushes may be defoliated. This is a serious pest of gooseberries and currants. **Control:** spray at the first signs of attack with Derris, Fenitrothion or Malathion, and repeat if necessary. Commercial growers may spray with Azinphos-methyl, a very poisonous chemical not available to amateur gardeners. **11 Grey mould** (see 4 above Botrytis). **12 Leaf spot** (see Blackcurrant 17 Leaf spot). **13 Magpie moth** *Abraxas grossulariata.* May attack young and established plants. Moths are black, white and yellow. Caterpillars are the looper type, with black and white markings and yellow stripes. **Symptoms:** caterpillars feed on the leaves in spring and early summer, eating holes out of them. This is no longer a serious pest. **Control:** spray with Carbaryl as the first flowers open, and again 3 weeks later. **14** Rust (see Blackcurrant 21 Rust). **15** Sawfly (see 10 above Gooseberry sawfly).

Grafting A method of propagation, used to produce new individuals quickly, and for various other reasons. A shoot, called a scion, of one plant is attached to the rooted part called the stock (see Stock) of another plant. This is a very old practice and there are a number of different methods, all requiring skill and care. Typical examples of plants produced through grafting are most fruit trees, for example, apples, pears, plums and peaches, where the stock exerts considerable influence on the vigour, fruiting habit and disease and pest resistance of the variety (scion). The technique generally employed is whip and tongue grafting. Other methods are saddle grafting for rhododendrons, root grafting for clematis, tree paeonies and gypsophila. Tomatoes are also grafted on pest and disease resistant root stocks. There are many other methods of grafting all employed basically for speed of propagation, and using the vigour or pest and disease resistance qualities of the stock to produce better results with the scion.

Grape hyacinth *Muscari botryoides* and varieties and other *Muscari* species and varieties. Hardy herbaceous perennials with bulbs. Height: small. Flowering time: spring. Flower colours: blue, purple-red, violet or white. Use: borders, naturalizing, rock gardens, pots in the cold greenhouse.

Pests and Diseases. 1 Black slime, caused by the fungus *Sclerotinia bulborum* (see Hyacinth 1 Black slime).

Grape vine *Vitus vinifera* varieties. Greenhouse and hardy (outside in sheltered positions in milder areas) deciduous shrub. Height: climbing. Flowering time: spring. Flowers: green, inconspicuous. Fruit: green, purple. Use: food crop grown for its fruit, also very decorative, greenhouse borders, commercial glasshouse borders, in the open in suitable places.

Pests and Diseases. 1 Botrytis (see 4 below Grey mould). **2** Brown scale *Parthenolecanium corni* (see Scale insects). May attack young and established plants. Attacks are worst in the spring. **Symptoms:** the reddish-brown scale insects encrust the main branches causing reduced growth and yields. **Control:** apply sprays of Diazinon or Malathion at 3-week intervals until the grape bunches are thinned. **3** Glasshouse red spider mite *Tetranychus urticae* (see Mites, Tomato 18 Glasshouse red spider mite). Note also that biological control (see Biological control) using Chilean predatory mite *Phytoseiulus persimilis* is very effective, provided that the predators are introduced as soon as damage is first seen. **4** Grey mould, caused by the fungus *Botrytis cinerea* (see Tomato 23 Grey mould). **5** Mealybugs (see Mealybugs). (a) Citrus mealybug *Planococcus citri*. May attack young and established plants. **Symptoms:** stunted growth; foliage marked by feeding; mealy, waxy mealybugs on the plants; small yellow elliptical eggs covered with mealy threads on the leaves. (b) Vine mealybug *Pseudococcus obscurus*. May attack young and established plants. **Symptoms:** mealy, waxy mealybugs on the plants; greyish sunken areas where mealybugs have fed. **Control:** spray with Diazinon or Malathion before the berries swell, otherwise the bloom will be spoiled. **6** Mildew (see 7 below Powdery mildew). **7** Powdery mildew, caused by the fungus *Uncinula necator* (see Mildew). May attack young and established plants. **Symptoms:** grey or purplish areas with powdery white fungal patches on the leaves and shoots; flowers and young berries may fall; berries may be cracked. **Control:** avoid overcrowding and checks to growth such as too dry an atmosphere or marked changes in temperature; ensure adequate ventilation without draughts; spray with Dinocap or sulphur (a) when laterals are 30–37 cm (12–15 in) long, (b) just before flowers open, (c) after berries have set; spray when temperatures are not too high. **8** Red spider (see 3 above Glasshouse red spider mite). **9** Scald (see Physiological disorders). May affect berries. **Symptoms:** berries have a scorched appearance. **Control:** avoid excessive sunlight and high temperatures early in the day. **10** Scale insects (see 2 above Brown scale). **11** Shanking (see Physiological disorders). May affect berries. **Symptoms:** berries at the tips of bunches develop spots and later wither; berry stalks turn a rusty colour; berries taste bitter. **Control:** ensure adequate moisture and correct feeding; do not allow too much fruit to develop, as shanking is worst on overladen vines. **12** Vine mealybug (see 5b above Vine mealybug). **13** Weevils, Wingless weevils *Otiorhynchus species*. May attack young and established plants. **Symptoms:** wilting foliage and damaged roots; U-shaped notches on the foliage; white, crescent-shaped legless grubs in the soil; adults which are nocturnal, climb foliage to feed. **Control:** drench soil and base of plant with gamma-HCH; apply gamma-HCH dust to the soil around the plants. Note that the Vine weevil *Otiorhynchus sulcatus* does not attack vines in Britain, though it does attack many other plants.

Grass   (see Lawns)
Greenfly   (see Aphis)
Grey mould   (see Botrytis)
Gypsophila   *Gypsophila elegans* and varieties, and *Gypsophila paniculata* and varieties. Hardy, herbaceous annual or perennial. Height: small to medium. Flowering time: summer. Flower colours: pink or white. Use: annual border, herbaceous border, garden and commercial outdoor cut flower, commercial glasshouse cut flower.
Pests and Diseases. 1 Stem rot, caused by the fungus *Sclerotinia serica*. May attack established plants. **Symptoms:** rotting at ground level causing plant to die back. **Control:** remove and destroy affected plants; plant new plants in different ground; if infection is suspected, drench plants with Benomyl.

H

Hawthorn   *Crataegus monogyna* Common hawthorn and varieties, *Crataegus oxyacantha* Hawthorn and varieties, and other *Crataegus* species. Hardy deciduous trees or shrubs. Height: trees, small; shrubs, large. Flowering time: early summer. Flower colours: pink, red or white. Berries: black, blue, crimson, red or yellow. Stems: spiny. Use: planted singly or in groups, hedges, hedgerows, windbreaks.
Pests and Diseases. 1 Mildew, caused by the fungus *Podosphaera oxyacantha* (see Mildew). May attack young and established plants. **Symptoms:** leaves and shoots covered with floury growth. This disease is most likely to be damaging on young trees. On large trees and hedges it is unsightly rather than damaging and any control is impractical. **Control:** use copper fungicides. **2** Rust, caused by the fungus *Gymnosporangium clavariaeforme* (see Rust). May attack young and established trees. **Symptoms:** raised orange spots on leaves and shoots, which cause malformation. **Control:** spray young trees with Maneb or Zineb. The other host is juniper, so

avoid growing juniper and hawthorn together (see Host).

Hazel Varieties of *Corylus avellana* and its forms, the Cobnut, Filbert and Barcelona nut. Hardy deciduous shrubs. Height: large. Flowering time: spring. Male flowers: decorative catkins. Fruit: a woody nut. Use: food crop, grown for the nuts; bushes or cordons. **Pests and Diseases. 1** Bud rot and twig canker, caused by *Gloeosporium* species of fungi. May attack established plants. **Symptoms:** buds rot and twigs die back. This disease causes serious restriction in cropping. **Control:** spray after picking but before leaf fall with mercury. **2** Caterpillars, Winter moth *Operophtera brumata*. **Symptoms** (see Apple 53 Winter moth). **Control:** spray with Carbaryl. **3** Nut drop, caused by the fungus *Sclerotinia fructigena*. May attack established plants. **Symptoms:** nuts fall. **Control:** spray with Benomyl, Bordeaux mixture, Captan, lime sulphur, Thiophanate-methyl or other recommended fungicide at regular intervals from bud-burst onwards. **4** Nut weevil *Curculio nucum*. May attack established plants. Brown weevils are seldom seen on the plants. **Symptoms:** eggs are laid in the young nuts, white larvae feed on the kernels and later overwinter in the soil; adult weevils emerge in May; considerable damage done in bad attacks. **Control:** spray with Carbaryl May/June.

Heat treatment This is treatment of plant material and seed by subjection to a specific temperature for a specific period of time. It is generally by immersion in hot water, a technique in which pests are killed while the plant material or seed is left unharmed. Temperatures are generally of a low order 42°C (110°F). But plants such as dahlias may also be grown at higher than normal temperatures 30°C (80–86°F) for a week or so to burn out any virus infection. Few heat treatments are readily adaptable for amateurs and are more often employed by specialists in plant or seed production.

Heath (see also Heath and Heather). *Erica carnea* and varieties, *Erica cinerea* Scotch or bell heather and varieties, *Erica hyemalis* South African heath, *Erica lusitanica* Portuguese heath and other *Erica* species and varieties. Hardy, greenhouse, evergreen shrubs and a few trees. Height: shrubs, prostrate, dwarf, medium, large; trees, small. Flowering time: throughout the year. Flower colours: lilac, pink, purple, red, white or yellowish-green. Foliage: may be copper, gold or reddish as well as pale and dark green. Use: borders, heather gardens, rock gardens, ground cover, greenhouse pots, borders, commercial glasshouse pot plants.

Pests and Diseases. 1 Heather beetle *Lochmaea suturalis*. May attack at all stages. **Symptoms:** leaves stripped and roots eaten. **Control:** drench with Malathion; Derris may also be useful. **2** Mildew, Powdery mildew, caused by the fungus *Erysiphe polygoni* (see Mildew). **3** Mussel scale (see 4 below Scale). **4** Scale, Mussel scale *Lepidosaphes ulmi* (see Scale insects). May attack young and established plants. Attacks are always worst in spring. **Symptoms:** stems encrusted with silvery grey insects, shaped like mussel shells; in severe attacks, leaves drop early in summer. **Control:** a drastic but effective treatment is a drench of tar oil in mid-winter, although this will kill the remaining foliage. **5** Wilt, caused by the fungus *Phytophthora cinnamomi* (see Wilt). May attack at all stages. **Symptoms:** first sign is a grey hue on the leaves, followed by wilting, usually on one side of the plant at first, then gradual dieback. **Control:** observe strict hygiene at propagating time; use sterilized composts, clean containers and clean water; propagate from healthy plants only. **6** Yellowing and distortion (see Physiological disorders). May occur on young and established plants. **Symptoms:** poor growth; yellow and distorted foliage. May be due to various causes, such as too much lime in the soil bringing the pH above 5 (unsuitable except for *Erica carnea* and its varieties), or by lack of certain organisms in the soil essential for the successful growth of heaths. **Control:** in both cases, use heavy dressings of sphagnum peat, taken from areas where heather has grown; apply flowers of sulphur to the soil in cases where the pH is extremely high.

Heath and Heather Heath is the name usually given to *Erica* species and varieties (see Heath). Heather is the name usually given to *Calluna vulgaris* and varieties (see Heather). The names are however interchangeable. *Erica cinerea*, for example, is known as Scotch or bell heather. Ling is a name given to *Calluna vulgaris*.

Heather (also called Ling) (see also Heath and Heather). *Calluna vulgaris* and varieties. Hardy, evergreen shrubs. Height: prostrate to dwarf. Flowering time: late summer to winter. Flower colours: lilac, pink, purple, red or white. Foliage: copper, gold or reddish as well as pale or dark green. Use: beds, borders, heather gardens, ground cover, *Calluna vulgaris alba* White heather, used as a cut flower. For **Pests and Diseases**, see Heath.

Helenium *Helenium autumnale* and varieties. Hardy herbaceous perennial. Height: small to medium. Flowering time: late summer, autumn. Flower colours: orange, yellow or parti-coloured. Use: herbaceous borders, cut flowers.

Pests and Diseases. 1 Leaf spot, caused by the fungus *Septoria helenii* (see Leaf spot). May attack established plants. **Symptoms:** brownish spots on leaves, which are unsightly rather than damaging. **Control:** spray with Benomyl.

Helichrysum *Helichrysum bracteatum* Everlasting flower and varieties. Half-hardy annual. Height: small to medium. Flowering time: summer. Flower colours: cream, brown, orange, pink, red, yellow or parti-coloured. Use: annual borders, grown in the garden or commercially to supply everlasting flowers. The flower heads dry naturally, becoming stiff and papery and retaining their colour. They will last for an indefinite length of time.

Pests and Diseases. 1 Downy mildew, caused by the fungus *Bremia lactucae*. May attack established plants. **Symptoms** and **Control** (see Lettuce 7 Downy mildew.)

Helleborus   (see Christmas rose)

Hippeastrum   (often called Amaryllis). *Hippeastrum* varieties grown are chiefly complex hybrids (see Hybrid). Greenhouse, more or less evergreen, herbaceous perennials with bulbs. Height: medium. Flowering time: varies according to variety, planting time and the way in which the bulbs have been treated, from winter to summer. Flowers: very large and trumpet-shaped. Flower colours: orange, pink, red, white or parti-coloured. Use: greenhouse and indoor pot plants.

Pests and Diseases. 1 Bulb scale mite *Steneotarsonemus laticeps* (see Bulb scale mite). May attack young and established plants. **Symptoms:** red streaks, on the bases of the leaves and stems which subsequently lengthen; poor growth; bulbs cut from nose to base show red marks on the scales near the neck. This is mainly a pest of forced narcissus but also attacks hippeastrum. **Control:** control measures are mainly beyond the amateur gardener but frequent white oil sprays will stop the mites from spreading. **2** Leaf scorch, Tip burn, caused by the fungus *Stagonospora curtisii*. May attack established plants. **Symptoms:** purplish to red spots or streaks of varying size on the leaves; spotting on flower stems and leaf stalks; infected bulbs small in size. **Control:** discard suspected bulbs or dip in a 0.8% formaldehyde solution for 1 hour; spray with a copper fungicide, Mancozeb commercially, or Zineb at shoot emergence, and at first signs of the disease; avoid over-watering. **3** Mealybugs *Planococcus* species (see Mealybug). **Symptoms:** poor growth; mealybugs present in the scales of dormant bulbs and in the leaves at the necks of bulbs. **Control:** drench with Demeton-S-methyl commercially or Malathion. **4** Tip burn (see 2 above Leaf scorch). **5** Virus, Tomato spotted wilt virus (see Virus diseases). **Symptoms:** pale blotches on the leaves which coalesce to form long streaks; leaves turn yellow and die; leaf margins may turn reddish; leaves may die. **Control:** destroy affected plants; spray to control insects which may spread the disease.

Holly *Ilex aquifolium* Common holly and varieties and other *Ilex* species and varieties. Hardy and greenhouse, deciduous and evergreen, trees and shrubs. Height: shrubs, prostrate to large; trees, small to large. Flowers: insignificant. Leaves: very decorative, shining, dark green, with toothed and sometimes spiny margins. Varieties of common holly may have attractive gold or silver variegated leaves. Berries: ornamental in autumn, red or crimson, occasionally yellow, according to sort. Use: planted singly or in groups, hedges, or uncommon hollies which are not hardy in the greenhouse.
Pests and Diseases. 1 Holly leaf miner *Phytomyza ilicis*. May attack established plants. **Symptoms:** blisters form on the leaves, where larvae have tunnelled. **Control:** on small trees, hand-pick affected leaves and burn them, or spray with Nicotine in spring; on large trees no control possible. **2** Holly leaf-tying moth *Rhopobota unipunctana*. May attack established plants. **Symptoms:** attacks on clipped hedges in the early summer; grey-green caterpillars tie the young leaves together with silk and strip the foliage, often causing browning of the remaining leaves. **Control:** spray with Carbaryl or Trichlorphon as soon as damage is seen; alternatively, carefully trim and collect and destroy the leaf trimmings.

Hollyhock *Althaea rosea* varieties. Not fully hardy herbaceous biennials. Height: large. Flowering time: summer. Flower colours: pink, red, white or yellow. Use: sheltered borders, borders next to walls.
Pests and Diseases. 1 Aphis, various Aphis species (see Aphis). **2** Cutworms (see Cutworms). **3** Rust, caused by the fungus *Puccinia malvacearum* (see Rust). May attack established plants. **Symptoms:** yellow oblong pustules on stems and leaves; round

orange-coloured pustules which eventually turn grey on fruits and calyces; premature leaf fall; greatly restricted growth. When this disease occurs there must be immediate action or the plants will be ruined. **Control:** at the first sign of attack, cut off affected parts and spray with a liquid copper fungicide, Maneb or Thiram. Some varieties are more resistant than others. **4** Slugs (see Slugs and snails).

Honesty *Lunaria annua* and forms. Annual often biennial. Height: medium. Flowering time: summer. Flower colours: purple or white. Foliage: variegated. Use: chiefly used for cutting. The seed heads are cleaned, leaving silvery, papery discs on the stiff branches, and are very decorative in dried bouquets.
Pests and Diseases. 1 Club root (see Brassicas 15 Club root or Finger and toe). Various other Brassica pests and diseases may attack honesty, which belongs to the same family.

Honey fungus (see Armillaria mellea)

Honeydew Certain pests, such as aphids, mealybugs, scale insects and whiteflies, excrete a sweet, sticky substance called honeydew. In bad pest attacks, honeydew coats leaves and other parts of plants. Various small fungi tend to grow on the honeydew and, being black in colour, they are called sooty moulds. Honeydew and sooty moulds restrict the amount of light reaching the leaves and so interfere with normal leaf functioning. Honeydew on plants attracts ants, which feed on it. Ants themselves are not particularly harmful to plants but they are unsightly and, as they move about, they may spread aphids and other pests.

Honeysuckle *Lonicera periclymenum* and varieties, and other *Lonicera* species and varieties. Hardy evergreen and deciduous shrubs. Height: shrubs, small to large and climbing. Flowering time: summer. Flower colours: purplish-red, yellow, yellowish-white or parti-coloured. Flowers: often very fragrant. Berries: black or red. Use: climbers, covering for arches, fences and walls, shrubs planted in groups, edging plants, hedges.
Pests and Diseases. 1 Leaf spot, caused by various fungi, or by weather damage. May attack established plants. **Symptoms:** spotted leaves, and, if a fungal disease is present, fluffy mould appears. Seldom damaging. **Control:** if fungal disease is present, spray with a liquid copper fungicide. **2** Mildew (see Mildew) caused by the fungus *Microsphaera alni* var. *lonicerae*. May attack established plants. **Symptoms:** fungal growth on foliage.

Control: if necessary, spray with fungicide. **3** Rust, caused by the fungus *Puccinia festucae* (see Rust). May attack established plants. **Symptoms:** orange pustules on undersides of leaves. **Control** (see Rust).

Hormone   A natural or artificial chemical which exerts an influence on various aspects of growth in man, animals and plants. In horticulture, certain hormones influence root development, and are used to treat cuttings. Others are used as weed-killers.

Host   The term used in biology for a plant or animal which supports as a parasite, or in some other way, another plant or animal. A plant affected by a pest or disease is said to be the host plant to that pest or disease.

House leek   *Sempervivum tectorum* and varieties, and other *Sempervivum* species and varieties. Hardy herbaceous perennials. Height: small. Flowering time: summer. Flower colours: pink, purple, red, yellow or white. Leaves: decorative, fleshy in close rosettes, may be covered with cobweb-like hairs. Plants may spread out widely, due to their habit of producing numerous young plantlets in the leaf axils.

Pests and Diseases. 1 Rust, caused by the fungus *Endophyllum sempervivi* (see Rust). May attack established plants. **Symptoms:** orange or red pustules on the leaves; affected plants are more upright in habit, with narrower leaves. **Control** (see Rust).

Hyacinth   Large-flowered and Roman hyacinths are derived from different *Hyacinthus* species. A few *Hyacinthus* species are grown. Hardy herbaceous perennials with bulbs. Height: small to medium. Flowering time: varies with method of growing from spring outdoors to winter to spring in the greenhouse or indoors. Flower colours: blue, cream, lilac, orange, pink, purple, red, white or yellow; very fragrant. Use: in the open, spring bedding, borders, containers, window boxes, in the greenhouse and indoors, pots and bowls.

Pests and Diseases. 1 Black slime, caused by the fungus *Sclerotinia bulborum*. **Symptoms:** yellow and collapsing leaves; decaying bulbs; bulb scales embedded with small black resting bodies of the fungus. **Control:** remove and destroy affected bulbs. Before planting hyacinths again, thoroughly sterilize border soil and use sterilized soil for pots and bowls (see Soil sterilization). **2** Bud drop and loose bud (see Physiological disorders). **Symptoms:** breaking off of flower buds as stems grow up. Usually due to starting to force bulbs at too early a stage and at too high a temperature. **Control:** follow correct forcing routine. **3** Floret withering (see Physiological disorders). **Symptoms:** withering and rotting of part of the flower spike. **Control:** follow correct rooting and forcing routine for indoor bulbs. **4** Flower discolouration (see Physiological disorders). **Symptoms:** browning or greening of part of the flower spike. **Control:** follow correct rooting and forcing routine for indoor bulbs. **5** Grey bulb rot, caused by the fungus *Sclerotium tuliparum*. **Symptoms:** failure of shoots to emerge, or shrivelling and dying of shoots shortly after emerging; soil sticking to the noses of the bulbs; spreading of rot from nose downwards, until bulbs are covered with fungal growth; bulbs embedded with small black or brown resting bodies of the fungus. **Control:** destroy diseased bulbs. Before planting hyacinths again, thoroughly sterilize the border soil, or incorporate Quintozene dust in the surface layers of the soil and use sterilized soil for pots and bowls (see Soil sterilization). **6** Penicillium bulb rot, caused by *Penicillium* species of fungus. **Symptoms:** soft brown rot of bulbs; blue-green mould coating bulbs. If infection is slight, and bulbs are grown under good conditions, this disease is not especially damaging. **Control:** discard soft bulbs. **7** Yellow disease *Xanthomonas hyacinthi* (see Bacterial diseases). **Symptoms:** streaks on the leaves, which turn yellow. If bulbs are cut, yellow slime oozes out. This disease should be confirmed by a plant pathologist. Ask the local horticultural adviser for advice. **Control:** destroy all diseased bulbs; treat outside ground with Formaldehyde (see Soil sterilization). In Holland, where the disease is severe, bulb raisers follow strict regulations to avoid export of infected bulbs.

Hybrid The term used for a plant raised from the crossing (see Cross-pollination) of 2 different species (see Species) or, occasionally, of 2 different genera (see Genus). The term bigeneric is used when 2 different genera are involved. Hybridization is a term used in a general way for the crossing of any 2 plants, though it often refers to controlled production by breeders of hybrids to supply new cultivars or varieties (see Cultivar, Variety). F1 hybrids (first filial generation hybrids) are plants resulting from a first cross between 2 distinctly different pure-bred strains (see Strain). F1 hybrids of various ornamental plants, and of agricultural and horticultural crop plants, have certain desirable characteristics, and also what is known as hybrid vigour, but seed taken from them will not necessarily give plants of the same quality. Second generation contrived crossing results in F2 hybrids, and so on.

Hybrid berries (see Blackberry)

Hybridization (see Hybrid)

Hydrangea *Hydrangea macrophylla* hybrids and varieties (see Hybrid), and other *Hydrangea* species and varieties. Hardy (except in cold areas), greenhouse, deciduous (one evergreen species) shrubs. Height: medium to large, climbing. Flowering time: in the open, summer to late summer; in the greenhouse, varies according to variety and method of growing from winter to spring. Flower colours: blue, lilac, pink, red or white. Use: in the open, beds, borders, containers, greenhouse and indoor pot plants, commercial glasshouse pot plants.

Pests and Diseases. 1 Aphis, various *Aphis* species (see Aphis). **2** Botrytis (see 9 below Grey mould). **3** Capsid, Common green capsid *Lygocoris pabulinus* (see Capsid bugs). May attack established plants. **Symptoms:** leaves of plants weakened or distorted. **Control:** spray with Dimethoate or Fenitrothion. **4** Chlorosis (see Physiological disorders). May occur on young and established plants. **Symptoms:** yellowing of the leaves; pale colouring of the flowers. **Control:** use suitable, lime-free composts for pot plants; in the open, water with a chelated iron product (see Sequestrols) to correct iron deficiency (see Mineral deficiencies). **5** Damping off, caused by the fungus *Rhizoctonia solani* (see Damping off). May occur on cuttings. **6** Discoloured leaves (see Physiological disorders). **Symptoms:** leaves discoloured, marked, or with brown margins. In the open, may be due to weather conditions and in the greenhouse or indoors, may be due to irregular temperatures or draughts. **Control:** in the open, plant in sheltered positions and give some protection in winter; in the greenhouse, maintain suitable temperatures and avoid having plants in draughts. **7** Eelworm (see 11 below Stem eelworm). **8** Glasshouse whitefly (see 12 below Whitefly). **9** Grey mould, caused by the fungus *Botrytis cinerea*. May attack established plants. **Symptoms:** rotting of flower heads. **Control** (see Botrytis). **10** Mildew, Powdery mildew, caused by the fungus *Oidium hortensiae* (see Mildew). May attack young and established plants. **Symptoms:** leaves covered with a mealy fungal growth. **Control:** spray with a copper fungicide; improve ventilation; do not over-water. **11** Stem eelworm

Ditylenchus dipsaci (see Eelworms). May attack at all stages. **Symptoms:** microscopic worms feeding inside the leaves, causing pale green swellings; later the leaves become crinkled, whilst the leaf stalks and stems shed their outer layers and become distorted and stunted; flowers are misshapen. **Control:** remove and burn affected plants; leave infested ground in the open clear of plants to starve the eelworms for three months; in the greenhouse, use sterilized composts; in mild attacks take new cuttings from vigorous shoots, provided there are no signs of eelworm damage. **12** Whitefly, Glasshouse whitefly *Trialeurodes vaporariorum.* May attack at all stages. **Symptoms:** masses of adult whiteflies on the plants; weakened and disfigured plants. **Control** (see Whitefly). **13** Yellowing of the leaves (see 4 above Chlorosis).

Hypericum   (see St John's wort)

I

Ilex   (see Holly)

Impatiens   (see Busy Lizzie)

In vitro culture   (see Meristem culture)

Insecticides   Often grouped under the more general heading of *pesticides*, insecticides are chemicals which kill insects. They can be divided into the contact type, effective against caterpillars and other defoliating pests, and the systemic type (see Systemic) which control sap-feeding insects such as aphids. Insecticides available to amateur gardeners in Britain are listed in the Appendix p. 245–54.

Iris   *Iris* species are divided into a number of groups according to the habit of growth. Many hybrids (see Hybrid) have been raised, and for convenience, the large bearded iris section is classified according to colour. Mainly hardy herbaceous perennials with fleshy roots and bulbs, corms or creeping rhizomatous rootstocks. Height: small to medium. Flowering time: early spring, summer in the open. Flower colours: very varied, shaded and splashed with contrasting colours and often parti-coloured; colours include amber, cream, blue,

blue/black, brown/red, grey, greenish, lilac, orange, pink, purple, white, yellow and parti-coloured. Use: iris borders and gardens, herbaceous borders, rock gardens, water gardens, pots in the cold greenhouse, commercial cut flowers in the open and in commercial glasshouses.

Pests and Diseases. 1 Aphis, various *Aphis* species (see Aphis). May attack at all stages. **Symptoms:** leaves covered with aphids. **Control:** dust with gamma-HCH or Nicotine dusts (see also 13 below Tulip bulb aphis). **2** Bacterial rot *Erwinia carotovora* (see Bacterial diseases). May attack rhizomes. **Symptoms:** leaf tips turn yellow and wither; soft wet rot of rhizomes. **Control:** lift rhizomes and cut out all diseased parts; replant in clean ground; dip material for propagation in corrosive sublimate (mercuric chloride, available from chemists only) 1:1000 for 10 minutes, handling with care as this is a very poisonous material; make sure that the ground is well drained and not short of phosphates. **3** Bryobia mites (see Mites, Primula 2 Bryobia mite). **4** Caterpillars (see Antirrhinum 7 Rosy rustic moth, Chrysanthemum 3 Caterpillars, Caterpillars). **5** Ink disease, caused by the fungus *Mystrosporium adustum*. May attack bulbs and established plants. **Symptoms:** dark blotches on leaves, which then turn yellow and die; discoloured outer scales of bulbs, as if ink-stained. **Control:** discard and burn severely affected bulbs; soak bulbs not badly affected in a 2% formalin solution for 1 hour immediately after lifting, and then dry and store until replanting time. **6** Iris sawfly *Rhadinoceraea micans*. May attack established plants. **Symptoms:** grey larvae feeding on the leaves; leaves bitten. **Control:** dust lightly with HCH or Nicotine dusts. **7** Leaf spot, caused by various species of fungi (see Leaf spot). May attack established plants. **Symptoms:** spotting and streaking of leaves which may be very disfiguring and damaging. **Control:** if attacks are severe, spray with Bordeaux mixture; cut off and burn old leaves. **8** Mites (see Mites, Primula 2 Bryobia mite). **9** Rosy rustic moth (see Antirrhinum 7 Rosy rustic moth). **10** Rotting of rhizomes, caused by the fungi *Botrytis cinerea, Sclerotinia* species (see Botrytis). May attack established plants. **Symptoms:** rhizomes rot and can easily be pulled out of the ground. **Control:** lift affected rhizomes, cut away rotted parts and treat with Benomyl; if possible, disinfect ground or replant on fresh ground; as this disease attacks through wounds, or attacks plants growing in very adverse conditions, take care to avoid mechanical damage to rhizomes, and improve condition of the soil. **11** Rust, caused by the fungus *Puccinia iridis* (see Rust). May attack established plants. **Symptoms:** orange-yellow pustules on the leaves during summer. **Control:** spray with Maneb (combined form), Mancozeb (commercially), Thiram or Zineb. **12** Sawfly (see 6 above Iris sawfly). **13** Tulip bulb aphis *Dysaphis tulipae*. May attack dormant tubers, worsening in the spring. **Symptoms:** brownish aphids cause severe stunting and distortion by feeding under the papery scales of tubers during the winter. **Control:** apply a drench of systemic pesticide in the autumn to protect the plants during winter (see Systemic). **14** Virus, Mosaic virus and Stripe virus (see Virus diseases). May attack established plants, chiefly bulbous irises. **Symptoms:** stunted growth; mottling of leaves; yellow streaking on flowers and stems; breaking of flowers (that is, flowers, or parts of flowers, may change colour or may be streaked with another colour). **Control:** destroy affected plants; spray to control insects which can spread virus diseases.

Ivy *Hedera helix* Common ivy and varieties and other *Hedera* species and varieties. Chiefly hardy, greenhouse, evergreen shrubs. Height: small to medium, climbing. Flowering time: summer. Flower colour: green. Berries: black. Foliage: ornamental, with great variation in form and leaf marking. Ivy in the climbing form does not flower, but when it has reached a height when no further climbing is possible, the leaves change shape and flowers and berries are produced. Cuttings rooted from flowering ivy make bushy shrubs. Use: covering walls and fences, ground cover on shaded banks and under trees, greenhouse border and pot plants, indoor pot plants, commercial glasshouse pot plants, bushes in borders, as edging to paths.

Pests and Diseases. 1 Leaf spot, caused by various species of fungi (see Leaf spot). May attack established plants. **Symptoms:** ash-grey spots with reddish margins on the leaves. It may be difficult to distinguish between leaf spots caused by fungal disease and leaf spots caused by scorching (see 2 below Scorching). **Control:** remove all spotted leaves, and if fungal infection is suspected, spray with Benomyl. **2** Scorching (see Physiological disorders). May occur on established plants. **Symptoms:** brown spots on the leaves; brown leaf margins or complete shrivelling of leaves. With indoor or greenhouse plants, this is usually due to irregular watering and sun scorch, especially where droplets of water dry out on the leaves. Cold draughts or warm air currents from for example, central heating, may also cause scorching. **Control:** remove badly affected leaves and improve general growing conditions; avoid drying out, bright sunshine and high temperatures. **3** Whitefly, Glasshouse whitefly *Trialeurodes vaporariorum* (see Whitefly). May attack established plants. Attacks greenhouse and indoor plants.

J

Japanese maple   (see Maple)

Juniper   *Juniperus communis* and varieties, and other *Juniperus* species and varieties. Hardy or almost hardy, evergreen coniferous trees and shrubs. Height: dwarf and prostrate shrubs to large trees. Foliage: decorative, may be blue grey or yellow, as well as green. Use: planted singly or in groups, ground cover, heather gardens, rock gardens.
Pests and Diseases. 1 Rust (see Hawthorn 2 Rust).

K

Kalanchoe   *Kalanchoe blossfeldiana* and varieties, and other *Kalanchoe* species. Greenhouse sub-shrubs with succulent leaves. Height: dwarf to small. Flowering time: winter, early spring. Flower colours: pink, red, white or yellow. Use: indoor and greenhouse pot plants.
Pests and Diseases. 1 Powdery mildew, caused by the fungus *Erysiphe polyphaga* (see Mildew). **2** Stem rot, caused by *Pythium*

and *Rhizoctonia* species of fungus. **Symptoms:** rotting of stems at soil level. **Control** (see Damping off, Root rot, Stem rot).

Kangaroo vine (see Cissus)

L

Larch *Larix decidua* and varieties, and other *Larix* species and varieties. Hardy deciduous, coniferous trees. Height: large. Foliage: decorative, pale green which colours to attractive browns and yellows in autumn. Growth: graceful. Flowering time: spring. Flowers: green, pink, red or yellow, followed by cones. Use: planted singly or in groups.

Pests and Diseases. 1 Canker, caused by the fungus *Trichoscyphella willkommii*. May attack young and established plants. **Symptoms:** under very wet conditions, canker wounds, with raised margins of white or orange fungal bodies. **Control:** little can be done with a large number of older trees; on a small scale, with young trees, cankers may be cut out, and the wounds treated with a mercury-based material such as Arbrex. **2** Leaf or needle fall, caused by the fungus *Meria laricis*. May attack young plants. Only younger trees seem to be affected. **Symptoms:** browning of leaves on upper branches. **Control:** use sulphur-based sprays or drenches for young trees. **3** Rust, caused by several different species of fungi (see Rust). May attack established plants. **Symptoms:** orange pustules on underside of leaves. **Control:** not a serious disease and control is neither necessary nor practicable on large trees.

Larix (see Larch)

Larkspur (see Delphinium)

Larva The name given to the immature stage of certain insects which, in their metamorphosis, pass from the egg, through larva and pupa, to the adult. It is usual to call a butterfly or moth larva a caterpillar, a sawfly larva a false caterpillar, a beetle or weevil larva a grub and a two-winged fly larva a maggot.

Lathyrus odoratus (see Sweet pea)

Laurel   (see Prunus)

Lavandula   (see Lavender)

Lavender   *Lavandula spica* and varieties. Hardy shrubs. Height: dwarf to small. Flowering time: summer to late summer. Flower colours: blue, lavender or white. Very fragrant. Foliage: grey-green and downy. Use: beds, borders, informal hedges, edging. Cut and dried and used for their perfume, grown commercially to supply oil for perfumery trade.

Pests and Diseases. 1 Aphis (see Aphis). May attack at all stages. **2** Cuckoo-spit bugs (see Cuckoo-spit bugs). **3** Phoma rot, caused by the fungus *Phoma lavandulae*. May attack established plants. **Symptoms:** yellowing and rotting of young shoots; eventual death of plant. **Control:** drench with liquid copper or commercially with Aaterra; plant new, healthy, young plants on fresh ground.

Lawns

Pests and Diseases. 1 Algae, Microscopic blue/green plants. Symptoms: slimy black growth on the grass, especially where it is water-logged, or where there is drip from trees. **Control:** improve drainage. **2** Ants, Common black ant *Lasius niger*, Yellow meadow ant *Lasius flavus*. **Symptoms:** ants nip off grass at the soil surface, but their obvious activity is the making of ant hills, which are unsightly and interfere with mowing. **Control:** use gamma-HCH dust, or liquid ant-killer near ant hills. **3** Birds. **Symptoms:** (a) birds tearing at the turf, usually a sign that leatherjackets or other grubs are present; (b) birds taking seeds from newly sown lawns. **Control:** (a) (see 13 below Leatherjackets, 20 below Wireworms); (b) use seed treated with protective chemicals. **4** Corticium disease, Red thread, caused by the fungus *Corticium fuciforme*. **Symptoms:** yellowish or, in severe attacks, brown areas of grass; if leaves are examined, tips will appear brown, and leaves will be seen to be dying; from the tips of dead leaves, reddish threads of the fungus grow out. This disease can be serious in summer and early autumn. **Control:** corticium is regarded as a sign of nitrogen shortage, so feed the lawn with fertilizer; treat affected areas with Benomyl, mercury, or Carbaryl/Quintozene. **5** Damping off, caused by various species of *Fusarium*, *Rhizoctonia* and *Pythium*. **Symptoms:** on newly sown lawns, young grass plants keel over at soil level; on established lawns, especially in winter and early spring, grass may turn yellow, particularly the meadow grass content, which is considerable in many lawns and sport turf areas;

grass may even turn mushy where there is much wear and tear on it. **Control:** in severe cases, use copper, Quintozene or Thiram preparations. **6 Dogs.** Bitches particularly may cause trouble. **Symptoms:** small burnt patches surrounded by dark green grass. **Control:** water the patches very thoroughly; keep dogs off the lawn. **7 Dog lichen.** Lichens are composite plants consisting of a fungus and an alga growing on decaying organic matter, and are a sign of poor lawn care. **Symptoms:** brown overlapping scales on established and especially fairly old lawn areas. Shade, bad drainage and starvation encourage their growth. **Control:** lightly scatter sulphate of iron on the lichen and, when it is blackened and dead, carefully rake it out or use a proprietary chemical killer; improve general growing conditions by spiking, feeding and possibly reseeding. **8 Dollar spot,** caused by the fungus *Sclerotinia homaecarpa.* **Symptoms:** small round yellow or brown patches, 5–8 cm (2–3 in) in diameter, appear, generally in late summer, and the disease is spread rapidly by foot or mower. This disease is less common than some of the other lawn diseases, but it may need treatment. It generally appears only in lawns with fine grass (creeping red fescue). **Control:** use the same chemicals as for Fusarium patch (see 11 below Fusarium patch). **9 Earthworms,** *Allolobophora* species. **Symptoms:** small piles or casts of fine, sticky soil produced in mild, moist weather in spring and autumn; the casts are unsightly and actually spoil the lawn, being so fine that they form a sort of mud when wet which may smother areas of grass; it is usually worse on heavier soils with a high organic content. **Control:** take proper care of the grass; if necessary, use a wormkiller; use Carbaryl, Chlordane, Derris or lead arsenate in autumn. **10 Fairy rings,** caused by *Marasmius oreades* and other species of fungi. **Symptoms:** symptoms vary according to the particular fungus present; there may be circles of toadstools and puffballs and a dark ring of grass, caused by the release of nitrogen by the fungus concerned; the grass in the centre of the ring may die off, or the ring may enlarge and eventually fade out, leaving the grass undamaged. **Control:** if treatment is obviously needed, holes should be made with a fork over the area of the rings and Formaldehyde injected into them; if this localized treatment is unsuccessful, remove turf and top-soil, without letting any fall on clean grass, and then fill in with fresh soil and re-turf or re-seed. **11 Fusarium patch,** Snow mould, caused by the fungus *Fusarium* *nivale*. **Symptoms:** brown or red/brown patches, quite small up to 30 cm (1 ft) in diameter, appear, usually in autumn though also at other times of the year; in moist conditions there is white mould on the patches, hence the name snow mould, although the muggy conditions under a blanket of snow encourage the growth of this fungus and this might equally be the source of the name. Melting snow predisposes grass to disease by depositing nitrogen in the soil, making grass growth soft. Although the grass leaves may be killed off, there can be good recovery from the roots. This is the most common of the lawn diseases. To establish if infection is by this fungus, cover a section of affected grass with a glass jam jar or glass dish. The snow mould will develop almost overnight if *Fusarium nivale* is present. **Control:** use a turf fungicide at once; a copper based fungicide, phenyl mercury acetoll (Calomel) or Quintozene may be applied as a preventative measure or on a curative basis; Benomyl has also been used in recent years with success. Intensively managed and heavily fed turf is much more likely to be attacked by this disease than lawns which are neglected. This is because of the softness and more disease-prone nature of the grass plants which are heavily fed. If attacks of Fusarium are neglected, turfing or re-seeding will be necessary. **12** Grubs (see 13 below Leatherjackets, 20 below Wireworms). **13** Leatherjackcts *Tipula* species (see Leatherjackets). **Symptoms:** irregular brown patches, which enlarge; birds searching in the lawn, tearing at the grass, and doing damage; if brown areas of grass come away easily by pulling, leatherjackets are probably present, and may be found in the ground; otherwise, to find out what pest is involved, lay down a few wet sacks overnight, and see what pests gather under them. **Control:** use gamma-HCH; feed the lawn in autumn. **14** Lichen (see 7 above Dog Lichen). **15** Mole *Talpa talpa*. **Symptoms:** a series of earth heaps made by moles tunnelling. Moles can do great damage to a lawn. **Control:** they are best dealt with by a professional mole catcher. **16** Moss. Not strictly a pest or disease. Moss is a very small non-flowering plant. Various species of moss affect lawns and one species or another can develop under almost any conditions. Moss grows best where the surface of the lawn is wet and greasy, that is, when drainage is restricted due to consolidation or other factors. Moss can thrive where grass cannot, so where grass is thin for any reason moss may invade the ground. Skinning the lawn by cutting too close, for example, can encourage the growth of moss. Lawns surrounded by trees are often subject to moss, since drips from the trees kill off the grass and encourage the moss to grow. But there are also types of moss which are able to compete with healthy growing grass. **Control:** acidity may encourage moss growth, so if the pH (see pH) is as low as 4.5–4.8, lime the soil, as a short-term measure; with ground limestone at 66 g per square metre (2 oz per square yard) on two or three occasions during the autumn or in early spring (the ideal pH for a quality lawn is 5–5.5); maintain correct general cultural conditions and ensure that moss control is part of the normal rountine of lawn care; for chemical control, use sulphate of iron and mercury in the form of Calomel, both in powder form; liquid solutions of mercury are available for use; follow directions for use carefully; moss killers may be used at any time of year, but preferably in March/April when the grass is growing strongly; in summer, moss killers may blacken the grass, so after moss is killed, it should be carefully raked out and the lawn should be fed; it may be necessary to re-seed bare patches. **17** Red thread (see 4 above Corticium disease). **18** Snow mould (see 11 above Fusarium patch). **19** Summer rust (see 4 above Corticium disease). **20** Wireworms (see Wireworms). Attacks on older grass are unimportant. **Symptoms:** some damage may be done in new grass, made worse by birds searching for the pests. **Control:** if necessary, use Chlordane; feed the lawn in autumn. **21** Worms (see 9 above Earthworms).

Leaf miners The larvae of various insects including beetles, moths, sawflies, two-winged flies and weevils are known as leaf

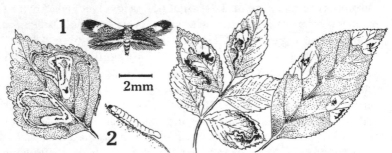

1 Moth   2 Larva and Leaf mines

miners, because they tunnel and feed between the two skins of the leaf, producing blotches or white wandering lines on the surface. A wide range of plants may be attacked, among them apple, brassicas, celery, chrysanthemum, rose and various trees and shrubs. The damage done reduces the effective functioning of the leaves, and so may check plant growth.

Control: use various insecticidal sprays such as Malathion, and spray *before* the plant becomes badly infested.

Leaf scorch Leaf scorch is a general term used to describe any sort of brown discolouration and dry marking on leaves. The cause may be physiological (see Physiological disorders), or there may be an actual pest or disease present. It is important to establish what is causing the scorching, so that a suitable control may be used.

Leaf spot Leaf spot is a general term used to describe any sort of spotting on leaves. The spots may be localized, separate areas of dead tissue, pale or dark, sometimes surrounded by pale green, yellow or reddish haloes, sometimes pitted or sunken with a definite rim, or the spots may have a water-soaked appearance. Spots may coalesce to form larger blotches or streaks of dead tissue and eventually the whole leaf may be destroyed. Leaf spots may be caused by mechanical, pest or spray damage, or by bacterial, fungal or virus diseases or by physiological disorders (see Bacterial diseases, Fungal diseases, Physiological disorders, Virus diseases). Many different plants may be affected. According to cause and circumstances, damage may be general or may be confined to certain kinds of plant, and may be mild, severe or fatal. Leaf spot symptoms may be associated with other symptoms, such as stem rot (see Stem rot). Control of leaf spot varies according to cause and circumstances.

Control: establish the cause of leaf spot so that appropriate measures can be taken; make sure that there is no risk of phytotoxicity, that is, that the plant is not liable to be damaged by a particular chemical used in a control material (information about phytotoxicity is often given with proprietary chemicals); follow directions carefully for using chemicals.

Leatherjackets Larvae of the crane flies or daddy longlegs flies, which are species of *Tipula*, are called leatherjackets. They are large, dark-grey, legless, wrinkled, and leathery, and live in the soil, especially in more moist soils. They may attack a wide range of plants. They feed at night, at or below soil level, cutting off stems in much the same way as cutworms (see Cutworms) and either cause plant to wilt, or destroy them.

Control: sterilize greenhouse border soil and composts, otherwise leatherjackets present in the soil may attack plants; use poison baits; spray with Fenitrothion or gamma-HCH (see Slugs and snails).

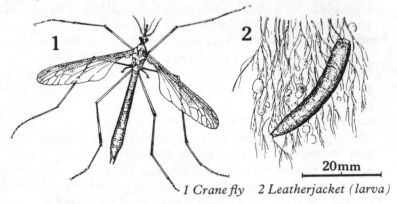

1 Crane fly   2 Leatherjacket (larva)

Leek   Varieties derived from *Allium porrum*. Hardy biennial, with thickened leaf bases. Height: medium. Use: food crop, grown for the blanched leaf bases.

Pests and Diseases. 1 Fusarium foot rot, caused by the fungus *Fusarium culmorum* (see Foot rot). May attack young and established plants. **Symptoms:** red colouration of plants at soil level followed by collapse and death. This is a disease of infected soil. **Control:** change the seed bed, or sterilize the soil with Basamid or a flame gun; practise rotation (see Rotation), which is especially necessary in this case. **2** Leek moth *Acrolepiosis assectella*. May attack young and established plants. **Symptoms:** moths, brownish-grey and 6 mm ($\frac{1}{4}$ in) long, overwinter in hedgerows or rubbish and eggs are laid in May on the bases of the plants; caterpillars hatch after 8 days and bore through leaves, causing considerable damage. In about 3 months, caterpillars are fully grown and become chrysalids inside cocoons which may be found among the dead leek leaves or near them. Moths appear about August, eggs are laid, and this second generation of moths overwinters. **Control:** use Derris or Fenitrothion. **3** Onion fly *Delia antiqua*. May attack young and established plants. **Symptoms:** the adult fly, which is like the common house fly, lays its eggs in stems or leaves at ground level in May and early June. Small larvae hatch out in 3 days time, eat into the base of bulbs and, when fully grown in about 3 weeks time, pupate in the soil. Adults

Larch **Canker** caused by the fungus *Trichoscyphella willkommii*, white and orange fungal bodies on bark (see page 122)

Lawns **Fairy rings** caused by the fungus *Marasmius oreades*, an enlarged ring which has left the grass inside undamaged (see page 124)

Lawns **Fusarium patch** caused by the fungus *Fusarium nivale*, red/brown patch on lawn (see page 124)

Lawns **Leatherjackets** *Tipula* species, leatherjackets in the ground near the surface (see page 125)

Lettuce **Big vein** transmitted by the fungus *Olpidium brassicae*, distorted and puckered outer leaf (see page 129)

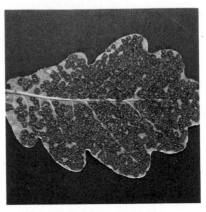

Oak **Oak gall wasp** damage caused by *Neuroterus* species, gall structures on the underside of a leaf (see page 148)

Onion **Onion fly** *Delia antiqua*, small larvae eating into base of bulb (see page 150)

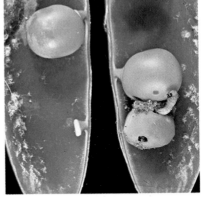

Pea **Pea moth** *Cydia nigricana*, caterpillars feeding on peas (see page 156)

emerge and lay eggs. There may be 3 generations in the season. Very damaging to the bulbs. **Control:** use Calomel dust around the plants in the rows and for seed dressing; buy treated seed. DDT or Dieldrin may be used by commercial growers. **4** Rust, caused by the fungus *Puccinia porri* (see Rust). May attack established plants. **Symptoms:** orange spots on the leaves, causing death of leaves; stunting of growth. **Control:** no really effective control, but spray with a copper fungicide or with Zineb which may reduce the disease. **5** White tip, caused by the fungus *Phytophthora porri*. May attack established plants. A very persistent disease especially in the autumn months. **Symptoms:** white tips on the leaves; complete rotting away of leaves, rendering the plants useless. **Control:** spray regularly with Bordeaux mixture.

Lettuce   Types and varieties derived from *Lactuca sativa*. Hardy annual. Height: small. Foliage: leaves make a rounded or long compact head, or a loose open head. Use: food crop, grown for the leafy heads. Grown in the open and in the greenhouse, in frames and other structures.

Pests and Diseases. 1 Aphis (see Aphis). (a) Glasshouse and potato aphid *Aulacorthum solani*, (b) Lettuce aphid *Nasonovia ribisnigri*. **Symptoms:** pale or dull green aphids about 2.5 mm ($\frac{1}{8}$ in) long on plants; curled or blistered leaves; stunted plants; honeydew and cast aphid skins on the leaves (see Honeydew). Some species penetrate to the hearts and spoil whole lettuces. **Control:** raise seedlings, as far as is possible, in an area free from aphids; inspect regularly for signs of aphid attack; spray with Dimethoate or Malathion; plant clean seedlings and continue to inspect for signs of aphids; use the above sprays on growing plants, giving thorough cover; follow directions for the use of sprays on edible crops carefully; keep down all weeds, since aphids may infest weeds. Commercial growers may use Demeton-S-methyl. Note that several species of aphids can transmit virus diseases (see 18 below Virus). (c) Lettuce root aphid *Pemphigus busarius*. May attack at all stages. **Symptoms:** clusters of white aphids attack roots and sufficient numbers may restrict growth. **Control:** dust soil with gamma-HCH or water with Malathion if on a small scale. Difficult to control if on a large scale. Diazinon is also recommended. **2** Big Vein (see 18 below Virus). **3** Black root rot, caused by the fungus *Thielaviopsis basicola*. May attack young and established plants. **Symptoms:** rotting roots; death of plants. **Control:** no control beyond removing and destroying diseased plants. This disease is not common so is not considered of great importance. **4** Collar rot, caused by the fungus *Rhizoctonia solani*. May attack established plants. **Symptoms:** brown markings on stems at soil level, followed by extensive rotting and collapse of plants. **Control:** sterilize border soil (see Soil sterilization); commercially, apply Quintozene to the surface. **5** Cutworms (see Cutworms). **6** Damping off, caused by *Pythium* species of fungus and *Rhizoctonia solani* (see Damping off). May attack at the seedling stage. **Symptoms:** collapse of seedlings at soil level. **Control:** for *Pythium*, remove affected seedlings and water with a copper fungicide or Thiram; for *Rhizoctonia*, incorporate Quintozene (commercially available) in the surface of the soil; always use sterilized soil for composts and clean containers. **7** Downy mildew, caused by the fungus *Bremia lactucae* (see Mildew). May attack young and established plants. **Symptoms:** pale green or yellow angular spots on the leaves; white fungal growth on the undersides of leaves. **Control:** use resistant varieties; pick off and burn initially infected leaves; spray with Thiram or Zineb; in the greenhouse, avoid humid conditions by ventilating and giving a little warmth. **8** Glasshouse and potato aphids (see 1 above Aphis). **9** Glasshouse symphilid *Scutigerella immaculata* (see Symphilids). **10** Grey mould, caused by the fungus *Botrytis cinerea* (see Botrytis). May attack at all stages. **Symptoms:** water-soaked spots on seedlings, which collapse; on older plants, reddish-brown spots on the stems at soil level; wet brown spots on leaves. **Control:** use resistant varieties; remove affected plants and destroy; spray with Benomyl, Thiophanate-methyl or Thiram; in the greenhouse, avoid humid atmosphere by ventilating and giving a little warmth, but do not allow plants to wilt; when sowing or planting, commercially rake Dicloran dust into surface soil. **11** Leaf spot, caused by the fungus *Pleospora herbarum*. May attack established plants. **Symptoms:** brown circular spots on the leaves which increase in size, coalesce and then dry up. Not a common or serious disease. **Control:** use fungicides such as Benomyl or Dichlofluanid which may give some control. **12** Root aphid (see 1 above Aphis). **13** Slaters (see Woodlice). **14** Slugs and snails (see Slugs and snails). **15** Springtails (see Springtails). **16** Symphilids (see Symphilids). **17** Tipburn (see Physiological disorders). May attack established plants. **Symptoms:** marginal scorching of the leaves. **Control:** use resistant varieties; avoid applying excessive amounts of fertilizer; ensure adequate soil moisture during hot weather. **18** Virus (see Virus diseases). (a) Big vein. A virus disease transmitted by a fungus *Olpidium brassicae*. May attack established plants. **Symptoms:** yellow or white vein banding makes veins appear prominent, especially at bases of outer leaves; distorted and puckered leaves. **Control:** efficient soil sterilization will kill the fungus which carries the virus (see Soil sterilization); raising plants in pots of sterilized compost will delay infection. (b) Lettuce mosaic or mosaic virus. May attack at all stages. **Symptoms:** seedlings grown from infected seed are stunted, with mottling of the young leaves; vein clearing, mottling, blistering and distortion on the leaves as the plants grow; plants are dwarfed. **Control:** spray with insecticide to control aphids which spread the disease; use mosaic-tested seed (containing less than 0.1% lettuce mosaic virus). **19** Woodlice (see Woodlice).

Ligustrum   (see Privet)

Lilac   *Syringa vulgaris* and varieties and other *Syringa* species and varieties. Hardy, deciduous shrubs and trees. Height: shrubs, medium; trees, small. Flowering time: early summer to late summer. Flower colours: cream, lilac, lilac-pink, purple and white. Mostly very fragrant. Use: planted singly or in groups, beds, gardens, grown in commercial glasshouses for cut flowers.

Pests and Diseases. 1 Bacterial blight, caused by *Pseudomonas syringae* (see Bacterial diseases). May attack established plants. **Symptoms:** water-soaked spots on the leaves; spots turn black and leaves shrivel. This is a typical bacterial disease mainly affecting forced lilacs grown in glasshouses. In these conditions, the lilacs are soft, due to high temperatures or feeding, and are liable to disease attack. **Control:** cut out all affected branches; reduce watering. **2** Bud rot, caused by the fungus *Phytophthora syringae*. May attack established plants. **Symptoms:** buds rot and brown spots appear on leaves. Forced lilacs grown in glasshouses are mainly affected. **Control:** avoid excess dampness; spray with various fungicides, but forced lilacs are very susceptible to leaf scorch. **3** Lilac leaf miner (see Privet 2 Lilac leaf miner). **4** Lilac wilt, caused by the fungus *Verticillium albo-atrum*. May attack established plants. **Symptoms:** typical wilt symptoms affecting one side of the plant, due to clogging of the conducting tissue (see Wilt). Mainly a disease of forced lilac. **Control:** destroy affected plants; always plant in clean compost or clean ground. **5** Silver leaf, caused by the fungus

Stereum purpureum. May attack established plants. **Symptoms:** leaves assume a silvery hue and branches die back. **Control:** (see Plum 15 Silver leaf).

Lily Lily varieties are complex hybrids derived from various *Lilium* species (see Hybrid). Lilies are put into a number of classes, depending on the species from which they have been derived. Some *Lilium* species are grown. Hardy, half-hardy or greenhouse, herbaceous perennials with bulbs. Height: medium to large. Flowering time: outside, summer; in the greenhouse, varies with the method of growing. Flower colours: cream, greenish-yellow, orange, pink, purple-red, white, yellow or parti-coloured, often conspicuously marked with contrasting colours. Flowers: often very fragrant. Use: beds, borders, herbaceous borders, woodland gardens, garden and commercial cut flowers in the open, commercial glasshouse cut flowers, greenhouse and commercial glasshouse pot plants.

Pests and Diseases. 1 Aphis, Tulip bulb aphis *Dysaphis tulipae* (see Aphis). **Symptoms:** grey aphids below outer skin of bulb; numbers increase rapidly after planting, causing stunted growth. **Control:** before planting, dip suspected bulbs for 15 minutes in a gamma-HCH solution (0.01%), or dust with gamma-HCH. **2** Leaf blight, caused by the fungi *Botrytis cinerea* and *Botrytis elliptica*. May attack at all stages. **Symptoms:** water-soaked or brownish spots on leaves, flower stalks and buds under humid conditions, and seedlings may damp off (see Damping off). **Control:** spray affected plants with a copper or sulphur fungicide, or use Benomyl; destroy old stems and leaves at the end of the season; discard diseased bulbs; sterilize soil before replanting (see Soil sterilization). **3** Leaf spot, caused by a *Cercosporella* species of fungus. May attack established plants. **Symptoms:** brown spots on leaves or stalks, which later develop a white mould. **Control:** spray with a liquid copper fungicide. **4** Lily thrips *Liothrips vaneeckei*. May attack established plants. **Symptoms:** minute, narrow insects feeding on bases of the leaves and bulbs; brown sunken areas where outer scales of the bulb have been attacked. **Control:** dust bulbs with gamma-HCH which is usually sufficient to prevent attack. **5** Root rot, caused by the fungus *Cylindrocarpon radicicola*. May attack established plants. **Symptoms:** gradual dieback of leaves; rotten roots. **Control:** plant bulbs in fresh ground which is in good condition, especially avoiding any risk of waterlogging. **6** Rust caused by a *Uromyces* species of fungus (see Rust). May attack established plants. **Symptoms:** brown pustules on the leaves. **Control:** pick off badly affected leaves; use a copper spray at the first sign of attack. **7** Storage rots, caused by various species of fungi. May attack stored bulbs. **Symptoms:** bulbs rot, and may be completely destroyed. **Control:** avoid any sort of physical damage to bulbs; store in cool, airy but frost-free conditions; avoid wet packing materials; wrap bulbs in waxed paper, which will help to keep them free of infection. **8** Thrips (see 4 above Lily thrips). **9** Virus mosaic and mottles, caused by various virus diseases (see Virus diseases). May attack established plants. **Symptoms:** distorted and mottled leaves; failure of flower petals to open – they remain held together at the tips. **Control:** destroy affected plants; spray with insecticide to control insects which spread virus diseases; propagate from bulbs which are known to be healthy, or propagate from seed.

Lily of the valley *Convallaria majalis* and varieties. Hardy herbaceous perennial. Height: small. Flowering time: spring. Flower: white, or tinged pink, sweetly scented. Use: beds, borders outdoors, woodland gardens, forced in pots, forced as commercial cut flowers.

Pests and Diseases. 1 Black lily of the valley, caused by the fungus *Sclerotium denigrans*. **Symptoms:** spots on buds which turn black. **Control:** outside, plant on fresh ground, using new, healthy plants; if bringing in forcing roots, store in sand and keep cool before forcing begins. **2** Grey mould, caused by the fungus *Botrytis paeoniae* (see Botrytis). **Symptoms:** distortion and eventual death of leaves; spotted flower buds. **Control:** if severe outside, cut out and burn all diseased parts; spray with Benomyl; spray early with Benomyl in following year: in greenhouse, remove and destroy badly affected plants; cut out and burn diseased parts; spray with Benomyl; avoid excess humidity during forcing.

Lime *Tilia europaea* common lime and other *Tilia* species and varieties. Hardy deciduous trees. Height: medium to large. Flowering time: summer. Flowers: small, greenish or yellowish-white, may be strongly scented. Use: planted singly, in groups or in avenues in large gardens, parks or open ground. Useful as street trees, as they stand pruning well. May be cut hard and trained to form screens.

Pests and Diseases. 1 Aphis, Lime leaf aphid *Eucallipterus tiliae* (see Aphis). May attack established plants. **Symptoms:** leaves covered with aphids, honeydew and black mould (see Honeydew). The common lime is most affected. *Tilia euchlora* is seldom affected. **Control:** spray with Malathion if trees are small enough. **2** Canker, caused by the fungus *Pyrenochaeta pubescens*. May attack established plants. **Symptoms:** sunken spots on the bark. **Control:** spray with a liquid copper fungicide in winter. **3** Leaf spot, caused by various species of fungi (see Leaf spot).

Ling   (see Heather)

Lobelia   *Lobelia cardinalis* and varieties, *Lobelia erinus* and varieties, and other *Lobelia* species and varieties. Not fully hardy, half hardy, greenhouse annuals and herbaceous perennials. Height: small to medium, trailing. Flowering time: summer. Flower colours: blue, red-purple, red, white or parti-coloured. Use: annual border, edging, summer bedding, window boxes, outdoor containers, outdoor hanging baskets.

Pests and Diseases. 1 Damping off, caused by various species of fungi, will attack germinated seedlings at early or later stages. **Symptoms:** plants keel over and die at compost level. **Control:** drench compost with liquid copper, or commercially with Milcol; avoid using newly heat-sterilized soil based composts or composts which have been stored for long periods, because of ammonia toxicity (see Soil sterilization). **2** Leaf blotch, caused by *Septoria* species of fungi. May attack established plants. **Symptoms:** blotches on leaves. Seldom really damaging. **Control:** spray with copper fungicide.

Loganberry   The original loganberry is said to have resulted from a natural cross between a variety of the raspberry *Rubus idaeus*, and a variety of the American blackberry *Rubus ursinus vitifolius*. There are a number of loganberry varieties. Hardy deciduous shrubs. Spreading habit of growth with strong arched shoots. Flowering time: spring. Fruit: dark red. Use: food crop grown for the fruit, trained to supports. For **Pests and Diseases**, see Blackberry.

Lonicera   (see Honeysuckle)

Lunaria   (see Honesty)

Lupin   *Lupinus polyphyllus* varieties, Tree lupin *Lupinus arboreus* varieties and Annual lupin *Lupinus luteus* varieties. Hardy annuals, herbaceous perennials and shrubs. Height: perennials, small to medium; shrubs, small to medium. Flowering time: summer. Flower colours: cream, blue, lilac, orange, pink, purple, red, white, yellow or parti-coloured. Use: annual borders, herbaceous borders, shrub borders, semi-wild gardens.

Pests and Diseases. 1 Brown spot, caused by the fungus *Ceratophorum setosum* and other species of fungi. May attack established plants. **Symptoms:** small dark spots on the leaves. **Control:** if troublesome, dust with a liquid copper fungicide. **2** Colour change. May affect established plants. **Symptoms:** lupins are said to 'revert', that is, the flower colour changes. This is due to cross-pollination, with the resulting seed falling to the ground, where it germinates and produces seedlings among the older plants. When the seedlings reach the flowering stage, their flowers may be of a different colour. With herbaceous lupins, blue tends to predominate. **Control:** always cut off all old flowering stems before the seed pods ripen. Note that, as lupins are legumes, like peas and beans, many of the troubles affecting peas and beans may also affect lupins (see Bean, Pea).

M

Mahonia *Mahonia aquifolium* and varieties, *Mahonia bealei*, *Mahonia japonica* and other *Mahonia* species. Hardy evergreen shrubs. Height: dwarf to large. Flowering time: winter, spring. Flower colour: yellow. Flowers may be fragrant. Berries: blue/black. Leaves: large, stiff and toothed. Use: shrub borders, woodland gardens.

Pests and Diseases. 1 Rust, caused by a *Puccinia* species of fungus (see Rust). **Symptoms:** brown spots with reddish margins on the undersides of the leaves and, less often, raised yellow spots. **Control:** spray with Maneb, Thiram or Zineb or commercially with Mancozeb; destroy very badly affected bushes.

Maple There are many maples, including the English maple *Acer campestre*, the Japanese maple *Acer palamatum* and varieties, the Norway maple *Acer platanoides* and varieties and the sycamore *Acer pseudoplatanus* and varieties. Mostly very hardy, mostly deciduous trees and shrubs, a few evergreens. Height: trees, small to large; shrubs, large. Foliage: variegated, well coloured in autumn. The decorative fruits of many maples are known as 'keys'. Use: planted singly or in groups, borders, avenues.

Pests and Diseases. 1 Aphis, Sycamore aphid *Drepanosiphum platanoidis* (see Aphis). May attack at all stages. **Symptoms:** large agile green aphids feeding on the undersides of the leaves and disfiguring them with honeydew (see Honeydew). **Control:** spray with Malathion in the summer if practical. **2** Coral spot (see Coral spot). **3** Leaf scorch (see **6** below Weather damage). **4** Mildew, Powdery mildew, caused by the fungus *Unicinula aceris* (see Mildew). May attack established plants. **Symptoms:** grey-white mould spreading over the leaves. Large, well-grown trees are seldom seriously affected, but trees growing in wet areas or in damp, overcrowded airless conditions may suffer. It is seldom practical to spray with chemicals except on small trees. **Control:** use copper spray, Dinocap or sulphur, or commercially Quinomethionate. **5** Tar spot, caused by the fungus *Rhytisma acerinum*. May attack established plants. **Symptoms:** large black spots on the leaves, which are disfiguring and cause premature leaf fall. **Control:** seldom necessary or practical. **6** Weather damage. May affect young and established plants, especially Japanese maples. **Symptoms:** leaves show withered edges or may be scorched and shrivelled. This may be due to cold winds or frost. **Control:** plant in suitable positions and provide shelter.

Marigold Two species share this common name, *Calendula* and *Tagetes*. Common or Scotch marigold *Calendula officinalis* and varieties. Hardy annual. Height: small to medium. Flowering time: summer. Flower colours: cream, orange or yellow. Use: annual border, summer bedding, window boxes and outdoor containers.

Pests and Diseases. 1 Leaf spot, caused by the fungus *Entyloma calendulae*. May attack established plants. **Symptoms:** white circular spots on leaves which turn brown. Not crippling. **Control:** spray with a copper fungicide. **2** Mildew, caused by the fungus *Sphaerotheca fuliginea*. May attack established plants. **Symptoms:** leaves covered with white mould. **Control:** spray with Dinocap.

African marigold *Tagetes erecta* varieties, French marigold *Tagetes patula* varieties and Striped Mexican marigold *Tagetes tenuifolia* varieties. Half-hardy annuals. Height: dwarf to medium.

Flowering time: summer. Flower colours: brownish-yellow, orange, red, yellow or parti-coloured. Foliage: strongly scented. Use: annual borders, summer bedding, outdoor containers, window boxes, short-term indoor and greenhouse pot plants.

Pests and Diseases. 1 Aphis, various *Aphis* species (see Aphis). **2** Damping off (see Damping off). **3** Foot rot, caused by a *Phytophthora* species of fungus (see Foot rot). May attack young and established plants. **Symptoms:** rotting of stems at ground level. **Control:** destroy plants; sow or plant in fresh ground. **4** Grey mould, caused by the fungus *Botrytis cinerea* (see Botrytis). May attack young and established plants especially in wet seasons, or when they have been damaged in some way. **Symptoms:** soft rot and grey mould on flowers and leaves. **Control:** spray with Benomyl. **5** Leaf spot, caused by bacterial infection or by weather damage (see Bacterial diseases). *Tagetes* are very easily damaged by frost. If spotting is seen in spring after frost, the plants will probably recover, but if spotting persists, bacterial infection may be present. **Control:** always use sterilized compost and clean containers for seedlings.

Marrow   (see Vegetable marrow)

Mealybugs   Mealybugs are closely related to scale insects (see Scale insects), but, unlike them, are active throughout life.

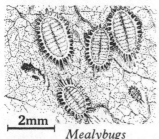

2mm

Mealybugs

Mealybugs are covered with a white, waxy substance, which gives them their name. Various species may be serious pests of a wide range of plants, chiefly in the greenhouse. Both young and adult mealybugs feed by sucking plant sap and so weakening plants. They excrete a sweet sticky substance called honeydew. Black moulds, known as sooty moulds, tend to grow on honeydew making plants unsightly and interfering with their normal growth. Mealybugs are more difficult to control than scale insects, because of their waxy impervious covering, and because of their habit of gathering inside curled leaves, beneath tree bark and in other hiding places. **Control:** regular routine spraying is necessary for effective control; use the various proprietary sprays available, but they should be used strictly in accordance with instructions. Some mealybug species (root mealybugs) attack plant roots, causing poor growth and yellowing of the leaves and if roots are examined, mealybugs may be seen. **Control:** use soil drenches of suitable insecticides.

Melon *Cucumis melo* and varieties. Greenhouse annual. Height: climbing. Flowering time: throughout the year, according to method of growing. Flower colour: yellow. Use: food crop grown for its fruit, in the greenhouse or commercial glasshouse.

Pests and Diseases. 1 Gummosis, caused by the fungus *Cladosporium cucumerinum*. May attack established plants. (See Cucumber 12 Gummosis). **2** Powdery mildew (see Mildew) caused by the fungus *Erysiphe cichoracearum*. (See Cucumber 14 Powdery mildew). For **Pests**, see Cucumber.

Meristem culture In meristem culture, the growing tips of shoots or meristems of plants considered free of virus and other diseases are propagated in seaweed extract solution (agar). This is a technique which needs specialized skill and care. In cell culture, also highly specialized, cells of plants are split up in extremely sophisticated machines and grown on in nutrient solutions.

Mice The short-tailed field mouse *Microtus hirtus* and the long-tailed field mouse *Apodemus sylvaticus* may be destructive by feeding on, for instance, bulbs and corms, fruits, roots, seeds and tree and shrub bark. They can also be troublesome in greenhouses. The house mouse *Mus musculus* may sometimes do damage in the garden or greenhouse by eating seeds, or by attacking certain plants, such as cucumbers, and eating the flowers and shoots. Occasionally voles or water rats may do damage in gardens.

Control: use traps or poison baits containing Coumatetralyl. Although this poison is more or less specific to rats and mice, it is best kept out of reach of pets, for example, by putting it in sections of drainage pipe, or having it safely covered in some way. The bait must be replenished for several days. Added sugar improves its attractiveness. A good hunting cat is also an effective control.

Michaelmas daisy In cultivation a number of *Aster* species, such as *Aster amellus*, *Aster novae-angliae*, *Aster novi-belgii* and others, have given rise to many varieties. A few *Aster* varieties may be hybrids. Hardy herbaceous perennials. Height: small to large. Flowering time: late summer, autumn. Flower colours: blue, pink, purple, red or white. Use: herbaceous border, borders, edging.

Pests and Diseases. 1 Eelworm (see Eelworm). **2** Mildew caused by the fungus *Erysiphe cichoracearum* (see Mildew). **Symptoms:**

white, powdery mould on stems and leaves; poor growth. **Control:** spray with a fungicide, if necessary; varieties vary in susceptibility, so discard those which are subject to mildew; many different types of plant are affected by this disease, so avoid other plants which are obviously susceptible. **3** Verticillium wilt, caused by the fungus *Verticillium vilmorinii*. **Symptoms:** wilting and drooping of leaves, which become distorted as the fungus clogs the conducting tissue on the stem; affected plants send up numerous small, new shoots from the base. **Control:** sterilize soil if practical (see Soil sterilization); otherwise, grow on fresh ground, and use healthy plants; if propagating from diseased plants, use suckers and do not divide crowns.

Mildew Mildew is a general name used for certain fungal diseases which are particularly noticeable on the leaves of plants and which may attack from the seedling stage onwards. It should be noted that the name mildew may be applied to various diseases, including botrytis (see Botrytis), but usually two main types of diseases are meant: downy mildew and powdery mildew. Different species of these diseases may attack a wide range of plants in the open and in the greenhouse. Downy mildew is more serious, as it lives inside the plant tissue, showing its presence by yellow discolouration of the leaves and general poor growth. Powdery mildew lives on the surface of the plant, giving it a white, powdery appearance. Both types are harmful, as they take food from the plant, and interfere with its normal functions. Correct growing conditions are very important if plants are to be kept as free as possible of mildew. For instance, in the open, mildew might be encouraged by factors such as too much water, too little water or cold weather. In the greenhouse factors might be irregular temperatures, inadequate ventilation, cold draughts or too much humidity.

Control, powdery mildew: cut off and destroy the parts of the plant affected, then use fungicidal sprays. **Control**, downy mildew: remove and destroy the plants affected and protect those that are left by spraying with fungicide; use any number of proprietary sprays which are available strictly according to instruction.

Millipedes Millipedes, species of *Tachypodojulus*, may be called black wireworms because they resemble wireworms, but are darker in colour (see Wireworms). There are other kinds, also slightly different in appearance. Millipedes move slowly, have rounded bodies with two pairs of legs to most of the body segments, and they curl up like coiled springs when disturbed. Millipedes may be confused with centipedes (see Centipedes). Millipedes lay eggs in the soil, and young and adult insects feed on decaying organic matter, but they may also attack living plants. Below ground, they

20mm

Millipede and millipede in curled up position

attack bulbs, tubers and roots, opening up the way to attack by diseases. They may also attack germinating seeds, especially peas and beans.

Control: this is difficult; use poison baits or traps (see Slugs and snails); dig Naphthalene (if available) into the soil; use insecticidal dusts based on gamma-HCH; sterilize compost or soil (see Soil sterilization).

Mineral deficiencies (or excess) When mineral deficiencies are suspected, first of all check that damage is not caused by exposure (wind), excess sunlight, lack of light, weather, pests, diseases, soil structure, high humidity, consolidation, excessive wetness, bad drainage, dryness, and root competition (from trees and hedges). Note that there could be several factors involved and that one issue could lead to another. The following account of plant nutrients is somewhat arbitrary but should serve to give initial guidance on whether mineral deficiency or excess could be involved. In extreme cases and when a considerable number of plants are involved, soil analysis should be undertaken. Note that symptoms vary with different plants. **1 Aluminium.** An exact role for this element is not known. Shortage is unlikely under most cultural systems (except hydroponics). Excesses are likely in very acid soils when insoluble aluminium salts are rendered soluble. Symptoms of deficiency: difficult to detect visually. Aluminium sulphate is used as a soil acidity chemical, especially for hydrangeas, to enhance colour. Symptoms of excess: failure or damping off (see Damping off) of lettuce and beetroot seedlings, coupled with root thickening and lack of root hairs. **Control:** for excess, avoid acidity for normal crop production, which causes excess of aluminium and other elements. 2 Ammonia. Has no role in plant growth apart from the nitrogen content. Shortage of ammonia is basically the same as nitrogen shortage (see 16 below Nitrogen). Excess is much more likely from newly sterilized soil or fresh farmyard manure. Symptoms of deficiency (see 16 below Nitrogen). Symptoms of excess: dark green spots or blotches on leaves, with very soft growth; scorched leaves, almost like frost or wind damage; seedlings are rendered very soft and are prone to damping off diseases (see Damping off). **Control:** for excess, avoid excess manure, especially fresh farmyard manure; flood soil to flush out excess and apply superphosphates of lime at normal application rates. 3 Arsenic. While arsenic has some role in plant nutrition, it is required in very small amounts and its function is conjectural. Shortage is unlikely to present any problems; excess is more likely. Symptoms of deficiency: unlikely to occur. Symptoms of excess: fertility of soil upset by killing off of micro-organisms at plant roots, causing plant death. **Control:** for excess, avoid the use of arsenical preparations. 4 Boron. Precise role in plant growth is not fully understood. Symptoms of deficiency: in tomatoes (see Tomatoes), stiff stems with dieback of terminal buds; leaves tend to be highly coloured, purple, brown or yellow; Raan Hollow Heart in turnips; Marsh spot of peas; tomato fruits may show death of cells below skin; in apples, damage to fruit is of a similar nature. Symptoms of excess: should seldom be a problem unless boron is over-applied (boron is a constituent of weedkillers, Bromacil for example). **Control:** for deficiency, use boron containing liquid feeds or apply boron very lightly to soil at 28 g per 17 square metres (1 oz per 20 square yards) and flush in with water. 5 Calcium (Lime). A major food which strengthens the cell wall formation of a plant. Deficiencies arise in acid soils in areas of high rainfall and excesses arise where limestone underlies top soil. Calcium also has an important role in sweetening the soil to neutralize acids created by micro-organism activity. Symptoms of deficiency: sourness of soil; poor growth; shoot dieback; hooking or angling of growth; undesirable chemicals such as aluminium or manganese can become soluble and be absorbed by plants, much to their detriment. Symptoms of excess: side effects include iron deficiency (see Sequestrols), and direct reaction with lime-hating plants such as rhododendrons, causing death in severe cases. **Control:** for shortage, lime the ground at rates according to pH figures (see pH);

for excess, use acid peat in quantity around lime-hating species, or apply Flowers of Sulphate at varying rates (see pH); acid forming fertilizer such as sulphate of ammonia is useful. **6** Chemical damage generally. Many pesticides (see Pesticides) can cause damage to plants if used incorrectly or at the wrong time. Symptoms can vary enormously and the first stage is to check back on what was used, when, and if it has been applied correctly. **7** Chlorine. This plays a very minor role in plant growth and it is an element unlikely to be in short supply under normal cultivation systems. Excess is a frequent problem due to widespread use of sodium chlorate weedkiller which is reduced to chlorine in soil. Symptoms of deficiency: seldom applicable. Symptoms of excess: paling or whitening of veins in leaves followed by greying and dropping, with stunted growth; roots of infected plants frequently very white. Sodium chlorate is a very soluble chemical and it can often wash into cropping areas. It has a devastating effect on many plants such as tomatoes and chrysanthemums in very small concentrations. Water supply can be suspect in some areas. **Control:** for excess, avoid indiscriminate use of sodium chlorate weedkiller. **8** Copper. Vital as a catalyst to assist chemical change in plant growth. More likely to be deficient under intensive or unusual cultural conditions in peaty land. Can be present in excess where industrial spoilage of land has occurred. Symptoms of deficiency: dieback of branch and disease of bark, especially apples and pears; stunting of growth or leaf rolling on soft crops such as tomatoes. Symptoms of excess: blue or black markings on stems of softer seedlings such as tomatoes. **Control:** copper deficiency can occur in very organic soils, so for deficiency improve these locally for apples or pears; avoid excess liming which can 'shut off' the copper; apply copper sulphate to soil at 8 g per square metre ($\frac{1}{4}$ oz per square yard); use sequestrol material for foliage. **9** Gas Poisoning. This can take many forms, from distorted growth to death by asphyxiation (lack of oxygen). Always inform the Gas Board immediately to carry out checks if a gas main is near and *do not attempt any personal investigation*. **10** Industrial Fumes. These can take many forms and will result in scorching (see 21 below Sulphur), discolouration, bleaching, poor growth and a wide range of other symptoms. This is a specialized matter, best taken up with a horticultural adviser. **11** Iron. A minor element which is a constituent of chlorophyl (the green leaf pigment). Present in most soils unless rendered unavailable by excess calcium or some other reason. Symptoms of deficiency: yellowing or whitening of leaf area *except* veins which remain green; mottling of leaves frequently occurs in young, quickly growing plants, but this symptom usually corrects itself; on apples and pears, rosettes of leaves can occur due to iron shortage. Symptoms of excess: seldom occurs, as excess iron is combined with other chemicals to form insoluble salts. **Control:** check pH; apply iron sequestrol at early stages. **12** Lead. Only a few cases of deficiency or excess symptoms recorded. Plants are more likely to act as carriers of lead derived from soil dressed with municipal compost or industrial waste. The effects of excess lead in human beings by absorbing lead from plants can be serious. **13** Magnesium. A major element which is an important constituent of chlorophyl. Shortages are more likely to occur in light, free-draining soils, especially under conditions of intensive culture under glass. Excess potash application can restrict uptake of magnesium. Symptoms of deficiency: interveinal yellowing of lower leaves, gradually moving up the plant; orange blotches develop; reduced growth and yield. Symptoms of excess: excess magnesium in soils by itself, or in combination with other elements, gives rise to soluble salt problems, which greatly restrict and damage plant growth (see Soluble Salts). **Control:** for deficiency, apply organic matter (farmyard manure if possible) to soil, especially if the soil is light; apply magnesium sulphate (epsom salts) to soil before starting crop, especially tomatoes; crops showing symptoms can be sprayed with Sequestrol or magnesium sulphate at 2% solution—1 kg per 25 litres (2 lb per 10 gallons)—plus spreading agent, starting sprays at an early stage of symptoms. **14** Manganese. A less important, but still essential, element in plant growth, which is also a vital constituent of chlorophyl. Manganese is not normally deficient in fertile soils under good conditions of drainage, but is deficient in soils of the lighter type subject to excessive rainfall. It is the toxic effects of manganese rendered soluble in acid or newly steam sterilized soils (see Tomatoes) which are likely to cause problems. Symptoms of deficiency: mottling and extreme wilting of leaves; a frequent malady in young, quickly growing plants in glasshouse culture. Symptoms of excess: leaves wither and droop, with dead patches. **Control:** check pH (see pH) and if too low, apply lime; apply manganese plus spreader if deficiency is bad. **15** Molybdenum. This has a vital role in all green plants. Deficiency is likely with brassicas, especially cauliflower which shows narrowing of leaves called Whiptail. **16** Nitrogen. This is one of the major plant foods with considerable influence on leaf growth. Being readily soluble it tends to be more deficient in lighter soils lacking in organic matter than in highly fertile loamy soils. Can also be deficient in highly organic wet soils such as peat. Symptoms of deficiency: smaller yellow green plants, often with reddish colouration. Symptoms of excess: soft dark green, much branching, unproductive growth; more leaves than flowers or fruits; physical damage to plants (see Soluble salts). **Control:** correct deficiency by applying nitrogen in either liquid or soluble form; balance excess by applying potash. **17** Oxygen. Vital for all living organisms and if either leaves or roots are deficient in oxygen, by for example flooding or gas contamination of soil, plants will show yellowing of leaves followed by death. **18** Phosphorus. A major plant food mainly concerned with root and fruit production. It tends to be deficient in cold, wet, acid soils or in certain regions of a specific soil type. Phosphate is not readily soluble. Symptoms of deficiency: dwarf, blue-green plants, often with reddish colouration; general lack of growth with poor roots, readily susceptible to disease. Symptoms of excess: seldom applicable, as excess phosphate is rendered unavailable. **Control:** for deficiency, apply phosphate in either solid or liquid form. **19** Potassium or Potash. A major plant food, its main role being to balance growth and produce good colour in flowers or fruit. Deficiency is more likely in lighter, free draining soil than in heavy clay soil. Symptoms of deficiency: scorching (browning) of leaf margins, coupled with very soft growth where there is plenty of nitrogen available. Symptoms of excess: scorching or browning of leaf tops and growing points; excessively curly leaves; shallow root systems; plants will have difficulty in extracting water from soil; in extreme conditions, plants, roots and stems will be chemically damaged (see Soluble salts). **Control:** check lime content of soil as too much lime can retard uptake of potash (see 5 above Calcium); for deficiency, apply potash in either solid (sulphate of potash) or liquid (a high potash type) form; take care not to over-apply as this will lead to soluble salt problems; avoid physical contact with the stems or roots of plants; for suspected excess, water liberally to flush out. **20** Soluble Salts (see Soluble salts). Excess of all elements, whether they be plant nutrients or weedkillers (for example sodium chlorate) result in toxic solutions which affect average plant growth. Excess salts, frequent in composts or highly managed greenhouse soils or growing media, can readily be checked by soil analysis. Symptoms of deficiency: excessively vegetative, soft, non-productive growth. Symptoms of excess: poor growth; burning or browning of leaf margins and growing tips; or death of plant. **Control:** for deficiency, apply fertilizer of the correct nutrient balance; flush out salts in the case of excess. **21** Sulphur. Shortages are unlikely under normal cultivation systems. The problem is more usually excess from polluted atmosphere. Symptoms of deficiency: long slender stems without side shoots, on soft crops such as tomatoes. Symptoms of excess: whitening or dead areas on leaves, especially of soft plants under glass. **Control:** check on fumes, especially from badly trimmed oil stoves; *always* ensure adequate ventilation; apply flowers of sulphur very lightly 34 g per square metre (1 oz per square yard). **22** Zinc. A very minor element in plant growth. Symptoms of deficiency: rosettes of small fruit on apples for example, with very narrow leaves. Symptoms of excess: very poor general growth of crops due to zinc derived from wire netting or overhead cables. **Control:** for suspected shortage, try a very light application (a few grams per square metre) of zinc sulphate; for excess, check source of zinc and report to appropriate body in the case of overhead wires; flooding helps excess.

Mint *Mentha rotundifolia* Apple mint, *Mentha spicata* Spear mint and other *Mentha* hybrids. Hardy herbaceous perennials. Height: medium. Foliage: aromatic. Use: herbs grown for flavouring, grown in the open and forced.

Pests and Diseases. 1 Cuckoo-spit bug (see Cuckoo-spit bug). **2** Eelworm (see 4 below). **3** Rust *Puccinia menthae* (see Rust). May attack established plants. May be very damaging on mint forced in the greenhouse. Symptoms: shoots and leaves thickened and distorted; yellow or brown pustules, which turn black later in the season, on the leaves. **Control:** spread straw on beds outside and burn over in autumn; plant runners taken from healthy plants only; lift dormant runners, wash in cold water, then soak in water at 44°C (112°F) for 10 minutes; wash at once in cold water and plant at once in fresh ground or use for forcing. This hot water treatment (see Hot water treatment) should clean the plants. **4** Strawberry eelworm *Aphelenchoides fragariae* (see Eelworms). May attack young and established plants. **Symptoms:** black patches on the foliage and distorted brittle shoots. **Control:** try to obtain healthy plants; treat runners for forcing with hot water treatment (see Hot water treatment); lift dormant runners, wash in cold water, soak in water at 46°C (115°F) for 10 minutes, wash at once in cold water and plant at once in fresh ground or use for forcing. Note that as mint is to be used in the kitchen, directions for spraying food crops should be carefully followed.

Mites   Mites are related to spiders, are very small or microscopic, and may be called red spider mites or spider mites. There are an extremely large number of different species, many of considerable economic importance, that is, they may cause damage serious enough to result in crop loss. Some mites are pests of a wide range of plants both in the open and in the greenhouse. They live and feed in slightly different ways, according to species, but typical symptoms are bleaching, mottling and yellowing of the leaves, due to all stages of the mite feeding on the undersides of leaves. Thick webs sheltering masses of young mites may be visible on leaves, flowers or fruit in severe attacks, and the mites may be seen with a magnifying glass.

Control: use one of the chemicals available (these are known as acaricides as well as insecticides or pesticides), but one problem is that certain strains (see Strain) of mite can build up resistance to a particular chemical, and so it is always important to vary the chemicals used in controlling mites; directions for the use of acaricides should be carefully followed; biological control (see Biological control) is also a possibility. The false red spider mites, such as those which attack cacti (see Cactus) are much smaller than red spider mites, though the damage they do when feeding is similar. (See also Bulb mite, Bulb scale mite.)

Molluscicides   These are pesticides made up as poison baits for the control of slugs and snails. See Appendix p. 245–54 for list of chemicals and poison baits.

Monk's hood   *Aconitum napellus* and varieties and other *Aconitum* species and varieties. Hardy herbaceous perennials, often with tuberous roots. Height: small to medium. Flowering time: summer. Flower colours: blue, purple, white or parti-coloured. Use: herbaceous borders, shaded borders, wild gardens. *Note* that all the aconites contain poisonous substances, so avoid planting within reach of cattle and avoid any confusion of the roots with root vegetables.

Pests and Diseases. 1 Aphis, various species, including the monk's hood aphid *Delphinobium junackianum* (see Aphis). May attack at all stages. Some varieties are much more subject to aphid attack than others, so might be discarded in the case of continual attacks. **Symptoms:** leaves and flower spikes covered with aphids and honeydew (see Honeydew), which weaken and disfigure the plants. **Control:** spray with insecticides (see Aphis). **2** Delphinium moth *Polychrysia moneta*. May attack at all stages. **Symptoms:** caterpillars, brown when young but later green with white and black lines, feeding on the buds, leaves, flowers and seed capsules, tying the leaves together with silk. **Control:** spray with Trichlorphon as soon as damage is seen. **3** Mildew, caused by the fungus *Erysiphe polygoni* (see Mildew). May attack established plants. **Symptoms:** mould on the leaves; restricted growth. **Control:** spray with a fungicide if thought necessary.

Mosquitoes These may be troublesome in water gardens and woodland gardens, but are annoying only to gardeners, not to plants.

Control: spray water with an insecticide, but not with Derris or with any other material harmful to fish.

Moths The caterpillars of most moths and some butterflies may be pests of a wide range of plants (see Caterpillars). Moth and butterfly pests are noted under the plants they attack.

Muscari (see Grape hyacinth)

Mustard and cress *Brassica nigra* Black mustard, *Brassica napus* Rape, and *Leptidium sativum* Cress. Annuals. Use: grown in the greenhouse in containers and cut at the seedling stage for salads and garnishing.

Pests and Diseases. 1 Damping off, caused by fungi of *Pythium* species, *Rhizoctonia solani*. **Symptoms:** dark, water-soaked marks on the stems at soil level; collapse and death of seedlings. **Control:** water with Cheshunt compound for pythium infection, and for rhizoctonia infection, commercially mix Quintozene with the compost; always use sterilized materials for compost and clean containers; avoid over-watering.

N

Nectarine Smooth-skinned form of peach (see Peach).
For **Pests and Diseases**, see Peach.

Nematicides These are chemicals mainly effective against eelworms. All of those so far discovered are extremely dangerous to
use and are not available to the amateur gardener.

O

Oak *Quercus robur* Common oak and its varieties, *Quercus ilex* Holm
or evergreen oak and other *Quercus* species and varieties. Hardy
deciduous and evergreen trees. Height: large. Flowering time:
spring. Fruits: decorative, woody and known as acorns. Use: singly
or in groups or avenues in large gardens, parks or open spaces.
Pests and Diseases. 1 Caterpillars of various moths, mainly
Tortrix viridana Green oak tortrix moth and *Operophtera brumata*
Winter moth. May attack young and established plants in spring.
Many moths lay their eggs on oak trees, but defoliation is usually
unimportant unless it occurs immediately after bud burst on young
trees. **Symptoms:** leaves tied together with silk. **Control:** spray
with either Carbaryl or Trichlorphon. **2** Gall wasps, *Diplolepis* and
Neuroterus species. May attack young and established plants in late
summer. **Symptoms:** a large number of small black wasps lay their
eggs in oak leaves and stems, causing a variety of gall structures such
as spongy spots on the undersides of leaves and oak-apples on the
stems. **Control:** usually unnecessary and not practicable but hand-
picking helps. **3** Leaf spot, caused by the fungus *Sclerotinia
candolleana* (see Leaf spot). **4** Mildew, caused by the fungus
Microsphaera alphitoides. May attack at all stages particularly

nursery seedlings and young trees. **Symptoms:** leaves covered with floury growth. Species of oak vary in susceptibility. **Control:** spray young seedlings in the nursery with Benomyl, Dinocap, lime sulphur, sulphur, or commercially with Quinomethionate.

Onion Varieties derived from *Allium cepa*, also other forms such as the Egyptian onion, the potato onion and the Welsh onion. Hardy biennials, with a bulb or bulblets. Height: medium. Basal leaves form a bulb, or bulblets form on stem. Use: food crop, grown for the ripened bulbs, or, as spring onions, pulled in the young stage.

Pests and Diseases. 1 Botrytis, caused by the fungus *Botrytis cinerea* (see Botrytis). May attack at all stages. **Symptoms:** white spotting of the leaves, which die back, checking growth; stored onions may turn mouldy and rot away, causing considerable loss. This disease may attack a wide range of different plants and is troublesome to onions in wet weather, especially to autumn-sown onions, as weather tends to be wet then. **Control:** spray with Benomyl; use adequate potash fertilizer; grow plants as well as possible, to avoid any check; pay attention to date of sowing in autumn, being careful not to sow too soon; store sound bulbs and inspect them regularly, getting rid of any diseased ones. Seed is now dressed with fungicides by most seed firms. **2** Downy mildew, caused by the fungus *Peronospora destructor* (see Mildew). May attack young and established plants. **Symptoms:** shrivelling of leaves from the tips downwards, some dying off completely; bulbs too are eventually affected. This disease may be troublesome in moist dull seasons and in wet areas. **Control:** at the first sign of attack, spray with Zineb or Bordeaux mixture. Some modern varieties are more resistant to this disease than others are, so might be tried. After an attack of downy mildew, ground should be rested for several years. Where a perennial onion bed is used, this must be changed, unless the soil can be sterilized with Basamid (see Soil sterilization). **3** Eelworm *Ditylenchus dipsaci* (see Eelworms). **Symptoms:** softening and distortion of leaves and necks. Symptoms might be confused with those of onion fly damage (see 6 below Onion fly), but in the case of onion fly, larvae will probably be found if several bulbs are cut down the centre. In the case of eelworm, the bulbs will look puffy inside. This eelworm may attack a wide range of crops and control is very difficult. **Control:** practise rotation (see Rotation), keeping the ground free of onions for several years; burn infested bulbs. Aldicarb granules may be incorporated when sowing by commercial growers only. **4** Mouldy nose (see 12 below White rot). **5** Neck rot, caused by the fungi *Botrytis allii, Botrytis byssoidea.* May attack seedlings of autumn-sown onions, but is more usually a disease of stored onions. May cause serious loss. **Symptoms:** collapse and death of seedlings; necks of stored bulbs turn black and later black spores can be seen on these areas. **Control:** it has been established that neck rot is seed-borne and seed can be bought which has been treated with a Benomyl and Thiram or other seed dressing (see Seed dressings). But it is very important to avoid any mechanical damage to onions, which may open the way for the disease to attack. Damage may be done when bending over the necks of the bulbs to encourage ripening before lifting. Hence the practice of pinching with the fingers before bending over the tops. When lifting, special care should be taken not to damage the bulbs and they should also be carefully handled when tying and storing. **6** Onion fly *Delia antiqua.* A serious and persistent pest of onions. May attack leeks and shallots. May attack young and established plants. **Symptoms:** the adult fly, which is like the common house fly in appearance, lays its eggs in stems or leaves at ground level in May and early June. Small larvae hatch out in 3 days, eat into the bases of bulbs and, when fully grown in about 3 weeks time, pupate in the soil. Adults emerge and lay eggs. There may be 3 generations in the season. Very damaging to the bulbs. **Control:** use an approved seed dressing; buy treated seed; use Calomel dust around the plants in the rows and for seed dressing. **7** Shanking, caused by a *Phytophthora* species of fungus. May attack established plants. **Symptoms:** yellowing of the leaves, followed by drying and shrivelling of bulbs. **Control:** difficult to control, but fortunately seldom occurs and probably more damaging to shallots. **8** Smudge, caused by the fungus *Colletotrichum circinans.* May attack established plants. **Symptoms:** black spots or smudges on the surface of the bulbs. Generally disfiguring rather than damaging. **Control:** store under cool, airy conditions. This disease also affects shallots. **9** Smut, caused by the fungus *Urocystis cepulae.* May be a serious disease. In certain areas there is legislation to control the sale of affected plants. May attack at all stages. **Symptoms:** dark lines at the growing points of affected seedlings; in older plants, dark blisters, which carry black spores on leaves and bulbs. **Control:** very difficult but rotation is essential (see Rotation). Failing this, soil sterilization with chemicals should be carried out (see Soil sterilization). **10** Soft rot *Erwinia carotovora* (see Bacterial diseases). This is a storage disease. **Symptoms:** areas on bulbs which look water-soaked, then turn soft and slimy and rot away. There may be considerable loss. **Control:** difficult to control; onions should be well grown and carefully lifted and stored, to avoid any possible damage; inspect stored bulbs regularly and remove any suspected of infection at once, as this disease may spread rapidly. It can affect many different kinds of plant in storage. **11** Stem eelworm (see 3 above Eelworm). **12** White rot, caused by the fungus *Sclerotium cepivorum*. May attack at all stages. May attack seedlings in the greenhouse, but more usually seen outside in May or June. **Symptoms:** yellowing and collapse of seedlings, older plants showing yellow leaves and later bases of bulbs covered with a white mould. Later the mould becomes felt-like and small black resting bodies of the fungus (sclerotia) may be seen embedded in it. These will remain in the soil and can start fresh attacks. **Control:** do not store diseased bulbs; rotation is essential and onions must be kept off infected ground for several years (see Rotation); alternatively, if possible, sterilize soil (see Soil sterilization); use Calomel seed dressing or buy treated seed (see Seed dressings), but seed must be sown in clean ground. **13** Virus, Yellow dwarf (see Virus diseases). May attack young and established plants. **Symptoms:** yellowing of leaves and dwarfing of plants. This disease is spread by certain sucking insects. **Control:** spray with insecticides.

Orchids General name given to a large number of different genera of herbaceous perennials (see Genus). There is considerable variation in mode of growth, those from the tropics being chiefly epiphytic (supported mechanically by other plants) while those from temperate regions are mainly terrestrial. A few are saprophytic (living wholly on dead animal and vegetable matter). There is great variation in flowering time, and in flower colour and form. The flowers may be remarkably beautiful and tend to be long-lived. Many orchids are difficult to cultivate. Use: pots and containers in the greenhouse, some grown in commercial glasshouses for cut flowers, hardy sorts in semi-wild gardens.

Pests and Diseases. 1 Aphis, Mottled arum aphis (see Cactus). **2** Citrus mealybug (see Camellia 4 Mealybug). **3** Fungus gnats, *Bradysia* species. May attack young plants. **Symptoms:** damaged roots; white larvae with black heads under pots and in compost.

Control: avoid waterlogging; treat infested compost with a drench of Diazinon. **4** Leaf and heart rot, caused by the fungus *Phytophthora cactorum*. May attack established plants. **Symptoms:** dark rot spreading quickly over leaves. **Control:** spray with a copper fungicide. **5** Leaf spot, caused by a number of different species of fungi. May attack at all stages. **Symptoms:** spotting and blotching of leaves. **Control:** spray with a general fungicide; grow plants as well as possible; if leaf spotting is noticed, isolate diseased plants at once. **6** Propagation fungus, caused by the fungus *Moniliopsis aderholdi*. May attack established plants. **Symptoms:** plants suddenly begin to rot, and no new shoots are formed. *Cypripedium* and *Odontoglossum* species are particularly susceptible. **Control:** spray with a general fungicide. **7** Scale insects, various species (see Scale insects). **Control:** use Malathion only. **8** Slaters or woodlice *Armadillidium* species (see Woodlice). **9** Springtails, *Collembola* species (see Springtails). **10** Virus, various virus diseases (see Virus diseases). **Symptoms:** mottling, ring spotting and streaking on leaves; colour breaking of the flower may occur, that is, flower may become parti-coloured. **Control:** destroy all diseased plants. **11** Wingless weevils *Otiorhynchus* species (see Grape vine 13 Weevils).

Osmosis All growing plants absorb water or moisture by a process called osmosis, where the stronger solution of salts in the plant cells, principally in the roots, draws in water from the weaker solution in the soil or growing medium. If an excess of salts should occur in the soil or growing medium, the reverse may happen, and the plant may be deprived of essential moisture, and growth would be restricted.

P

Paeony *Paeonia lactiflora* Paeony rose and varieties, *Paeonia officinalis* Paeony rose and varieties, *Paeonia suffruticosa* Moutan paeony and varieties and other Paeonia species and varieties. Mostly hardy herbaceous perennials and shrubs. Height: herbaceous perennials, medium; shrubs, small. Flowering time: early summer. Flower colours: pink, purple-red, red, white, yellow or parti-coloured. May have ornamental seed heads, and foliage which colours well in autumn.

Pests and Diseases. 1 Blight, caused by the fungus *Botrytis paeoniae*. May attack young and established plants. **Symptoms:** soft brown areas at the base of shoots or leaves, which then die; flower buds may turn soft and fail to open. In wet areas, may be very serious, causing dieback of plants. **Control:** dust with copper fungicide, or spray several times in summer at intervals with Benomyl; in autumn, after cutting down and burning old stems, carefully scrape away soil from the crowns of affected plants, and replace it with clean soil from elsewhere; lime the ground, if it is acid; if crowns are to be divided, select healthy plants, lift carefully in autumn or spring, dip in disinfectant, and replant in fresh ground. **2** Leaf spot, caused by the fungus *Septoria paeoniae*. May attack young and established plants. **Symptoms:** spots on leaves. Seldom damaging. **Control:** spray with a copper fungicide. **3** Phytophthora, caused by a *Phytophthora* species of fungus. May attack established plants. **Symptoms:** dark, leathery markings on leaves and stems. **Control:** spray with a copper fungicide.

Palms   A general term applied to a large number of different genera (see Genus) of woody perennial plants. There is considerable variation in the size and form of the leaves. Half-hardy and greenhouse perennials. Height: dwarf to small. Foliage: ornamental. Use: borders or pots in the greenhouse, pots indoors, containers in the open in summer, summer bedding.

Pests and Diseases. 1 Leaf spot and wilting of leaf tips. May be caused by various species of fungi or may be physiological (see Physiological disorders). May attack young and established plants. **Symptoms:** brown spots and streaks on the leaves; browning and wilting of leaves at the tips. Palms suffer from a wide variety of leaf spots and leaf wilting. Leaf spots may also be caused by dryness of the soil or of the atmosphere, by overwatering of the soil, by water droplets on the leaves drying in hot sunshine, or by fungal infection. **Control:** improve growing conditions; take special care to avoid either overwatering or an over-dry atmosphere; if careful examination shows that fungal disease is present, spray with a fungicide.

Parsley   Varieties derived from *Petroselinum crispum*. Hardy

Parsley 1 Carrot fly

biennial. Height: small. Foliage: plain, curled or crested. Use: a herb, grown for the leaves, used for flavouring or garnishing.

Pests and Diseases. 1 Carrot fly (see Carrot 5 Carrot fly). May attack parsley. **2** Leaf spot, caused by the fungus *Septoria petroselini* (see Leaf spot). May attack established plants. **Symptoms:** brown spots, later turning white, on the leaves. Like celery late blight or leaf spot, this is a seed-borne disease. **Control** (see Celery 5 Late blight). **3** Violet root rot, caused by the fungus *Helicobasidium purpureum* (see Carrot 12 Violet root rot). May also attack parsley.

Parsnip Varieties derived from *Peucedanum sativum*. Hardy biennial, with fleshy roots. Height: medium. Roots: long, thickened and tapered. Colour: creamy-yellow. Use: food crop, grown for its roots.

Pests and Diseases. 1 Bacterial soft rot *Erwinia carotovora* (see Bacterial diseases, Carrot 10 Soft rot). **2** Canker (see Physiological disorders). Various pests and also species of bacteria and fungi are associated with this disorder. May attack established plants. **Symptoms:** cracking of the upper part of roots, with dark areas and rotting. This trouble is more usual where there is excess nitrogen in the soil, and especially where farmyard manure has been recently applied, and where there is a deficiency of lime in the soil. **Control:** there is no real control, but the damage may be reduced or prevented by correct manuring and liming, and by adequate pest control. Varieties vary in susceptibility, and some which show resistance are available. **3** Carrot fly (see Carrot 5 Carrot fly). **4** Mildew, caused by the fungus *Erysiphe umbelliferarum* (see Mildew). May attack established plants. **Symptoms:** powdery white growth on leaves. **Control:** none necessary. This disease seldom does much damage. It may be associated with sclerotinia rot (see 5 below Sclerotinia rot). **5** Sclerotinia rot, caused by the fungus *Sclerotinia sclerotiorum*. Chiefly attacks roots in store, but may also attack plants. **Symptoms:** rotting of roots and plants. This disease may attack many different kinds of plants, in store or while growing, and the resting bodies of the fungus (sclerotia) persist in the soil and can start infection in subsequent crops. **Control:** practise rotation (see Rotation); inspect stored vegetables and bulbs, removing and destroying any that are diseased. **6** Soft rot (see 1 above Bacterial soft rot).

Pea Varieties derived from *Pisum sativum*. Hardy annual. Height: climbing, small to large. Flowering time: summer. Flower colours:

red or white. Pods: short. Use: food crop, grown for its seed.

Pests and Diseases. 1 Aphis, Pea Aphid *Acyrthosiphon pisum* (see Aphis). May attack young and established plants. **Symptoms:** distorted leaves and lack of growth; aphids visible on plants. **Control:** spray with Dimethoate, Fenitrothion, Malathion or Oxydemeton-methyl. For commercial growers the following are listed in the *Ministry of Agriculture Agricultural Chemicals Approval Scheme* booklet: Azinphos-methyl mixtures, Demephion, Demeton-S-methyl, Dichlorvos, Dimethoate, Fenitrothion, Formothion, Malathion, Mevinphos, Oxydemeton-methyl, Phosphamidon, and Thiometon. **2** Birds (see Birds). **3** Black root rot, caused by the fungus *Thielaviopsis basicola*. May attack young plants. **Symptoms:** new tap roots turn black; wilting and death of plants. This disease is common in colder areas and in soils with a high lime content. **Control:** practise rotation and use seed dressed with Captan/Thiram (see Rotation, Seed dressings). Some nurserymen use seed dressings as a regular precaution. Acid-forming fertilizers such as sulphate of ammonia or possibly flowers of sulphur may be used to lower the soil pH but such action should be taken only as a last resort (see pH). **4** Damping off, caused by *Pythium* species of fungi. May attack seed. **Symptoms:** rotting of peas in the ground and failure to germinate. Usually occurs with autumn or early spring sowings. **Control:** can be almost entirely eliminated by using a Captan/Thiram seed dressing (see Seed dressings). **5** Downy mildew, caused by the fungus *Peronospora viciae* (see Mildew). May attack young and established plants. **Symptoms:** dark mould on the underside of the leaves, which may spread to the pods. This tends to be a disease of wet seasons. Not too damaging unless the pods are attacked, when there may be considerable loss. **Control:** remove and destroy badly affected plants; use Zineb, but this is not recommended for use on peas in the *Ministry of Agriculture Agricultural Chemicals Approval Scheme* booklet. **6** Eelworm (see Eelworm, Potato 20 Golden potato cyst eelworm). May attack established plants. **Symptoms:** yellowing of leaves; pinhead-sized cysts on the roots, white at first, later turning yellow. **Control:** very difficult to control. Feeding with a nitrogen fertilizer may help plants to grow out of an attack. After an attack, the use of Basamid (Dazomet) soil sterilant is likely to be helpful, but peas should be kept off affected ground for a number of years by practising a long rotation (see Soil sterilization,

Rotation). Note that plants are often affected by both eelworm and Fusarium foot rot (see 8 below Fusarium foot rot). **7** Foot rot (see 8 below Fusarium foot rot). **8** Fusarium foot rot, caused by *Fusarium solani* and other *Fusarium* species of fungi. May attack at all stages. **Symptoms:** turning brown of lower leaves; thinning and darkening of stems at ground level; black rotting on roots. This disease causes a great deal of trouble in peas, both culinary and decorative. Frequently occurs in soils of good fertility, especially those well supplied with organic matter, when temperatures are high. Round-seeded varieties are said to be more susceptible to this disease than wrinkle-seeded ones. **Control:** buy dressed seed or use Drazoxolon commercially or Thiram seed dressings (see Seed dressings); practise rotation and in extreme cases, chemical soil sterilization (see Rotation, Soil sterilization). **9** Golden cyst (see 6 above Eelworm). **10** Leaf and pod spot, caused by the fungi *Asochyta pisi, Asochyta pinodella, Mycosphaerella pinodes*. May attack established plants. **Symptoms:** round spots on the leaves, brown or black according to the species of fungus involved, later becoming sunken. Pods and seeds may be attacked and infected seeds shrink badly, and may act as a source of infection if sown. **Control:** use Thiram seed dressings and practise rotation (see Seed dressings, Rotation). **11** Marsh spot (see Mineral deficiencies). **Symptoms:** brown spots inside the peas. If affected seeds are sown, they may fail to germinate. This disorder is due to manganese deficiency in the soil. **Control:** apply manganese sulphate at 70 g per square metre (2 oz per square yard) and if on a large scale, at 630 kg per hectare (5 cwt per acre). **12** Mildew, Powdery mildew, caused by the fungus *Erysiphe polygoni* (see Mildew). May attack established plants. **Symptoms:** leaves and pods covered with white fluffy mould later in the season, especially in dry spells of weather. **Control:** if attack is severe, spray with Dinocap. Some varieties, such as Kelvedon Wonder, show resistance. **13** Mosaic (see 23 below Virus). **14** Pea aphid (see 1 above Aphis). **15** Pea and bean weevil (see Bean, broad 3 Bean weevils). **16** Pea beetles (see Bean, broad 2 Bean beetles). **17** Pea midge *Contarinia pisi*. May attack established plants. **Symptoms:** tiny flies causing damage to flowers at the opening stage. **Control:** spray with Dimethoate or Fenitrothion. Commercial growers may also use Azinphos-methyl mixtures. **18** Pea moth *Cydia nigricana*. **Symptoms:** maggoty peas. Eggs are laid on the plants from June to August and the black-backed caterpillars bore into the pods and feed on the peas. **Control:** use the following chemicals: Azinphos-methyl mixtures and Tetrachlorvinphos commercially, Carbaryl or alternatively Fenitrothion. **19** Pea thrips *Kakothrips pisivorus*. May attack established plants. **Symptoms:** reduced growth and badly deformed pods. Small black thrips lay eggs in May and June, and the larvae feed on flowers, leaves and pods, giving leaves and pods a silvery marking. **Control:** use Azinphos-methyl mixtures commercially, Dimethoate and Fenitrothion. **20** Powdery mildew (see 12 above Mildew). **21** Root rot, caused by the fungus *Aphanomyces euteiches*. May attack seedlings and young plants. **Symptoms:** plants grow to a few shoots, then die. **Control:** use seed dressings (see Seed dressings). This trouble is often bad in cold seasons and on wet soils. **22** Thrips (see 19 above Pea thrips). **23** Virus, Mosaic virus (see Virus diseases). May attack young and established plants. **Symptoms:** mottled leaves and, in bad attacks, deformed pods. This virus is spread by various sucking pests. **Control:** use appropriate insecticide for pest involved. **24** Weevil (see Bean, Broad 3 Bean weevils).

Peach Varieties derived from *Prunus persica*. Hardy, but usually grown in the greenhouse or against walls in the open, deciduous trees. Flowering time: spring. Flower colour: pink. Fruit: yellow, flushed red, downy-skinned. Use: food crop, grown for the fruit, trained against walls in the greenhouse, against walls and sometimes as trees in the open.

Pests and Diseases. 1 Aphis, Black peach aphid *Brachycaudus persicae*, Peach aphid *Brachycaudus schwartzi*, Peach-potato aphid *Myzus persicae*, Mealy peach aphid *Hyalopterus amygdalus* (see Aphis). May attack young and established plants. **Symptoms:** aphids of various colours and honeydew on the leaves and shoots (see Honeydew); severe leaf curl and leaf fall on young shoots. **Control:** spray with Oxydemeton-methyl, Malathion or Derris in mid-May and 3 weeks later; vary spray material used frequently. **2** Blossom wilt (see 3 below Brown rot and blossom wilt). **3** Brown rot and blossom wilt, caused by the fungi *Sclerotinia fructigena* and *Sclerotinia laxa*. May attack established plants. **Symptoms and Control** (see Apple 16 Brown rot and blossom wilt). **4** Brown scale *Parthenolecanium corni* (see Scale insects). May attack established plants. **Symptoms:** shoots stunted and covered with brown scales which do not move but increase until after a few years they may cover the bark completely; growth is weakened. **Control:** spray with Diazinon or Malathion in the growing season; spray with tar oil winter wash in the dormant season. **5** Eelworm (see 12 below Walnut root lesion eelworm). **6** Fruit tree red spider mite *Panonychus ulmi* (see Mites). May attack young and established plants. **Symptoms:** mites may be seen with a hand lens; mites feed on the leaves, giving them a silvery appearance and greatly reducing their efficiency. Control is made difficult by the fact that strains of the mites develop, resistant to the various chemicals used (see Strain). **Control:** spray with Dimethoate or Malathion after blossoming. **7** Glasshouse red spider mite *Tetranychus urticae* (see Mites). May attack young and established plants in the greenhouse. **Symptoms** (see Tomato 18 Glasshouse red spider mite). **Control:** spray with Demeton-S-methyl commercially, Dimethoate or Derris in mid-May and again 3 weeks later. **8** Leaf curl, caused by the fungus *Taphrina deformans*. May attack young and established plants. **Symptoms:** leaves distorted, thickened and yellow, with tinges of red, which gradually become redder until the leaves wither and fall. This disease is common outdoors but infrequent in the greenhouse. **Control:** pick off infected leaves; gather fallen leaves and burn all of them; spray before bud burst and after leaf fall with a copper fungicide or lime sulphur. **9** Powdery mildew, caused by the fungus *Sphaerotheca pannosa* var. *persicae*. May attack established plants. **Symptoms:** infected shoots stunted, with narrow leaves; powdery growth on the leaves and shoots. This disease attacks plants in the greenhouse, though old, neglected, congested outdoor plants may sometimes be affected. **Control:** spray with Dinocap, Quinomethionate commercially, or a proprietary sulphur fungicide at regular intervals. **10** Red spider mite (see 6 above Fruit tree red spider mite, 7 above Glasshouse red spider mite). **11** Scale (see 4 above Brown scale). **12** Walnut root lesion eelworm *Pratylenchus vulnus*. May attack young plants. **Symptoms:** stunted plants with poor root systems; dead leaves. This pest may also attack apricots. **Control:** remove and destroy affected plants; sterilize the soil with D-D commercially, or use fresh soil (see Soil sterilization); buy clean healthy young plants from a reliable source.

Pear Varieties derived from *Pyrus communis*. Hardy deciduous trees. Height: medium. Flowering time: spring. Flower colour: white. Fruit: greenish-yellow or yellowish-brown. Use: food crop, grown for the fruit, may be grown in pots in the greenhouse on a small scale. Where plants in the greenhouse may be attacked, this is noted.

Pests and Diseases. 1 Aphis, Pear and bedstraw aphid *Dysaphis pyri*, Pear and coltsfoot aphid *Anuraphis farfarae*. May attack young and established plants. **Symptoms:** aphids and honeydew on the leaves (see Honeydew). Aphids are not usually a serious pest of pears. **Control:** if bad attacks occur, spray with Derris, Dimethoate, Malathion or other suitable insecticide. May attack plants in the greenhouse. **2** Apple and pear bryobia mite *Bryobia rubrioculus* (see Apple 3 Apple and pear bryobia mite). This pest is becoming of increasing importance to pears. It migrates from grass, clover and other weeds, where there are 5 generations. There are 2 generations on pears. **Control:** as well as spraying, it is useful to keep trunks free of grass and weeds to a radius of 60 cm (2 ft). **3** Bacterial blossom blight *Pseudomonas* species (see Bacterial diseases). May attack established plants. **Symptoms:** withering of infected flowers, followed by spotting of fruit. **Control:** none so far recommended. **4** Blossom wilt (see Apple 16 Brown rot and blossom wilt). **5** Bryobia mites (see 2 above Apple and pear bryobia mite). **6** Canker, caused by the fungus *Nectria galligena* (see Apple 4 Apple canker). May also attack plants in the greenhouse. **7** Capsid, Common green capsid *Lygocoris pabulinus* (see Apple 5 Apple capsid). **8** Caterpillars (see Apple 25 Codling moth, Apple 35 and 51 Tortrix moths, Apple 53 Winter moth, 15 below Pear sawfly). **9** Fire blight *Erwinia amylovora* (see Bacterial diseases). **Symptoms:** shoots wilt and die; blossom wilts; by mid- or late-season, withered brown or blackened leaves hang on the branches; cankers appear on branches and trunk; old cankers ooze in spring and are a source of infection; tree is finally killed. This is a very serious disease of pears, which may also attack apples, though it is not as yet, considered of great concern on Apples (see Apple 30 Fire blight). **Control:** there is no effective chemical control. This disease is notifiable, that is, it must be reported to the Ministry of Agriculture. **10** Fruit tree red spider mite *Panonychus ulmi* (see Mites). **Symptoms** (see Apple 33 Fruit tree red spider mite). This pest is becoming more common on pears than formerly. **Control:** spray with Chlorpyrifos commercially, Dicofol (in combined form) or Dimethoate in mid- to late-June. May attack plants in the greenhouse. **11** Leaf and bud mite *Phyllocoptes schlechtendali* (see Apple 38 Leaf and bud mite).

12 Pear leaf blister mite *Phytoptus pyri*. May attack established plants. **Symptoms:** greenish-yellow blisters on the leaves which turn red then brown and the whole leaf often turns black and falls; fruitlets may also be attacked. **Control:** spray with Carbaryl at bud burst. **13** Pear leaf midge *Dasineura pyri*. May attack young and established plants. **Symptoms:** small white active insects feed on the young leaves, making them roll inwards. **Control:** spray with Carbaryl, Fenitrothion or Malathion in mid-June and repeat twice within a week, and again after an interval of 3 weeks if damage persists. **14** Pear midge *Contarinia pyrivora*. May attack established plants. **Symptoms:** small white larvae feeding inside pear fruitlets, causing them to turn blackened and misshapen; affected fruitlets usually fall. **Control:** clear up and burn affected fruitlets; spray in the following season with Carbaryl at white bud stage. **15** Pear sawfly *Hoplocampa brevis*. May attack established plants. **Symptoms:** caterpillars are white with black head and tail markings; damage to fruit similar to that caused by Apple sawfly (see Apple 8 Apple sawfly), with ribbon-like markings and holes in the fruit. **Control:** spray with gamma-HCH 7 days after approximately 80% of the petals have fallen. **16** Pear scab, caused by the fungus *Venturia pirina*. May attack established plants. **Symptoms:** (see Apple 9 Apple scab). This disease is closely related to Apple scab. **Control:** best controlled by going on to a regular spraying programme. May also attack plants in the greenhouse. **17** Pear sucker *Psila pyricola*. May attack established plants. **Symptoms:** developing buds turn brown due to pale aphid-like insects feeding; buds are destroyed and later leaves are distorted and covered with whitish honeydew and moulds (see Honeydew); plants are weakened. **Control:** in spring and summer, spray with Chlorpyrifos commercially, Dimethoate, Fenitrothion or Malathion; in the dormant season, use a tar oil wash on a mild day to kill overwintering adults. **18** Sawfly (see 15 above Pear sawfly). **19** Scab (see 16 above Pear scab). **20** Tortrix moth (see Apple 35 and 51 Tortrix moths). **21** Winter moth (see Apple 53 Winter moth).

Pelargonium *Pelargonium* varieties are complex hybrids derived from various *Pelargonium* species (see Hybrids). For convenience, varieties may be put in the following sections: ivy-leaved, regal, species of scented-leaved and zonal. Pelargoniums are popularly but incorrectly called geraniums (see Geranium). Greenhouse,

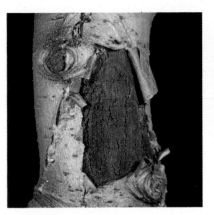

Pear **Fire blight** *Erwinia amylovora*, canker on trunk (see page 159)

Pine **Blister rust** *Cronartium ribicola*, burst blisters releasing spores (see page 167)

Pine weevil *Hylobius abietis* (see page 167)

Plum **Plum fruit moth** *Cydia funebrana*, hole in fruit made by the red caterpillar which feeds near the stone (see page 169)

Potato **Blight** caused by the fungus *Phytophthora infestans*, premature destruction of leaves (see page 173)

Potato **Colorado beetle** *Leptinotarsa decemlineata*, beetles with their characteristic yellow stripes, a notifiable pest (see page 174)

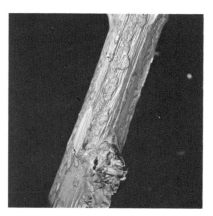

Raspberry **Cane spot** caused by the fungus *Elsinoe veneta*, pale and sunken spots on cane which later crack (see page 185)

Rose aphid *Microsiphum rosae*, yellow-green aphids clustering on leaf (see page 193)

half-hardy, some hardy outside in mild areas, herbaceous perennials which may be woody-stemmed. Height: small to medium, tall against walls, trailing. Flowering time: throughout the year, according to the method of growing. Flower colours: orange, pink, purple, red, white or parti-coloured. Foliage: succulent, attractively variegated, or strongly scented. Use: greenhouse and indoor pot plant, hanging basket plants, greenhouse wall-trained plant, wall plant in the open in mild areas, summer bedding, borders, containers or hanging baskets outside in summer, window boxes.

Pests and Diseases. 1 Aphis, various species (see Aphis). **2** Bacterial leaf spot and stem rot *Xanthomonas pelargonii*. A bacterial disease. May attack at all stages. **Symptoms:** slightly sunken, water-soaked spots on the leaves which later turn brown; water-soaked lesions near the stem apex which spread slowly upwards and downwards, and later turn dark; finally the stems become dry and shrunken; cuttings infected at the base appear similar to cuttings infected with blackleg (see 3 below Blackleg). **Control:** check that plants are healthy before taking cuttings from them; note that ivy-leaved pelargoniums can carry this disease without showing any symptoms of it, so keep these apart from other varieties. (For general hygiene, see Damping off, Foot rot.) **3** Blackleg, caused by *Pythium* species of fungi. May attack at all stages. **Symptoms:** soft black rot of stems; rotted roots. **Control:** take cuttings from healthy plants only. (For general hygiene, see Damping off, Foot rot.) **4** Black root rot, caused by the fungus *Thielaviopsis basicola*. May attack at all stages. **Symptoms:** plants turn yellowish and wilt; black rotting on roots. **Control:** water soil with Benomyl (see also Damping off, Foot rot). **5** Caterpillars, of various species of moth. **Symptoms:** caterpillars on plants; leaves eaten. **Control:** use Derris dust. **6** Colour tinting of leaves (see Physiological disorders). May occur on young and established plants. **Symptoms:** leaves, especially lower leaves, turn blue or red, usually red. This disorder is due to cold. Plants usually grow out of it when heat levels improve. Some varieties are more affected than others. **Control:** where possible, improve growing conditions. **7** Glasshouse whitefly *Trialeurodes vaporariorum* (see Whitefly). **8** Grey mould, caused by the fungus *Botrytis cinerea* (see Botrytis). May attack at all stages. **Symptoms:** soft rot and grey mould on leaves, flowers or stems. **Control:** spray with Benomyl or

Dichlofluanid; reduce humidity. **9** Leaf curl (see 14a below Leaf curl virus). **10** Leafy gall (see Sweet pea 4 Leafy gall). **11** *Oedema* (see Physiological disorders). May attack established plants. **Symptoms:** small, scab-like swellings on the leaves. The swellings are caused by the atmosphere being so moist that the plants cannot give off water by transpiration. **Control:** reduce humidity; maintain suitable night and day temperatures; avoid marked differences in temperature. **12** Ringspot (see 14b below Ringspot). **13** Rust, caused by the fungus *Puccinia pelargonii-zonalis* (see Rust). May attack young and established plants. **Symptoms:** concentric brown rings mainly on the lower surfaces of the leaves; yellowish areas on the upper surfaces. **Control:** pick off and burn affected leaves; destroy severely affected plants; spray at least fortnightly with Maneb, Thiram or Zineb or commercially with Mancozeb; grow plants on the dry side; free cuttings from infection by allowing them to wilt and then put them in a polythene bag at 38°C (100°F) for 48 hours. It is best when preparing cuttings to trim them cleanly just below a node or leaf joint. **14** Virus, various virus diseases (see Virus diseases). (a) Leaf curl virus. May attack young and established plants. **Symptoms:** small pale green or yellowish spots, sometimes becoming star-shaped, on leaves; leaves more or less crinkled or distorted. Symptoms are more pronounced in spring. **Control:** go over plants and remove and destroy any showing symptoms; take cuttings from healthy plants only. (b) Ringspot, Pelargonium ringspot virus. **Symptoms:** light green spots and ring spots on the leaves; stunted plants. **Control:** as for leaf curl virus (a) above. **15** Whitefly (see 7 above Glasshouse whitefly). **16** Yellowing of leaves (see Physiological disorders). **Symptoms:** lower leaves turn yellow, then drop off. Some varieties are more susceptible than others. May be due to overwatering. **Control:** allow the plants to dry out, if the soil seems to be too wet; overfeeding may sometimes have the same effect, so check feeding.

Peperomia *Peperomia* species and varieties. Greenhouse annuals and perennials. Height: small, creeping or trailing. Flowering time; spring, summer. Flowers: insignificant, or slender white catkins. Leaves: fleshy, often very decorative in shape, colour and variegation. Use: greenhouse border, hanging basket and pot plant, indoor pot plant.

Pests and Diseases. 1 Aphis (see Aphis). **2** Leaf spot (see 4 below

Virus). **3** Leaves soft and pale (see Physiological disorders). May occur in young and established plants. Usually due to overwatering, especially in winter. **Control:** reduce water. **4** Virus, Leaf spot or Ring spot, possibly caused by Cucumber mosaic virus (see Virus diseases). **Symptoms:** stunted plants; distorted leaves, with pale ring spots; spotted areas may die. **Control:** discard affected plants; use insecticidal sprays to control aphids, if present, as they may spread the disease.

Pepper Varieties of *Capsicum annuum* including the common capsicum. The large pods are often used green, and the chilli, with small pods, may be used green in pickles or for making vinegar, or red, to supply cayenne pepper. A number of varieties are grown as ornamental plants. Greenhouse, annual or herbaceous perennial. Height: medium. Flowering time: summer. Flowers: small, greenish-white. Fruit: pods vary in size and shape, and may be cream, red or yellow in colour when ripe. Use: food crop, grown for the fruit pods, also grown as ornamental pot plants for the sake of the decorative pods.

Pests and Diseases. Peppers are subject to many of the troubles which affect tomatoes (see Tomato). Glasshouse red spider mite, may be troublesome (see Tomato 18 Glasshouse red spider mite).

Pesticides This is the universally accepted term for all pest and disease controlling chemicals, whether they are insecticides (see Insecticides), acaricides (see Acaricides), nematicides (see Nematicides), molluscicides (see Molluscicides), poison baits (see Poison baits) or fungicides (see Fungicides). Seed dressings (see Seed dressings) are intended to protect against pests generally whereas soil sterilants (see Soil sterilization), as well as being pesticides, control weeds and can also be regarded as herbicides. When using pesticides, first and foremost, read the directions on the label. Most pesticides available to the amateur gardener are comparatively safe but it is good practice to wear rubber or plastic gloves, rubber boots and waterproof clothing, especially when handling the concentrated material. Measure out the quantity of materials accurately. Avoid the use of teaspoons, drink bottles and other utensils which may be used for eating or drinking. Be careful when applying pesticides to plants not mentioned on the label because of the risk of plant scorch (phytotoxicity). This applies particularly to succulents, ferns and variegated plants. If in doubt, test and spray one or two plants and observe them for several days before spraying the entire batch. Choose suitable weather to avoid drift when spraying outdoors and apply at the time of year recommended on the label. With edible crops, observe the stated harvest interval. Certain materials should not be used where a later crop is likely to be tainted with residual amounts of pesticide (for example, gamma-HCH should not be used where potatoes or carrots are to be grown in the following two years).

Petunia *Petunia* varieties and hybrids derived from *Petunia integrifolia* and *Petunia nyctaginiflora* (see Hybrid). Half-hardy and greenhouse annuals. Height: small. Flowering time: summer. Flower colours: cream, blue, lilac, orange, pink, purple, red, white, yellow or parti-coloured. Use: annual borders, summer bedding, containers, hanging baskets in the open, window boxes, greenhouse pot plant, greenhouse hanging baskets, short-term indoor pot plant. **Pests and Diseases. 1** Aphis (see Aphis). **2** Foot rot, caused by the fungus *Phytophthora cryptogea* (see Foot rot). **3** Leaf spot, caused by the fungi *Phyllosticta petuniae* and *Ramularia petuniae*. May attack established plants. **Symptoms:** large, round, brownish spots on the leaves. **Control:** pick off and burn affected leaves; not a common trouble, but if severe, spray with a copper fungicide. **4** Leafy growth, with no flowers (see Physiological disorders). May occur in established plants. **Symptoms:** leafy, bushy plants, which fail to flower, or produce few flowers. This is a nutritional problem. **Control:** give potash feeding; do not plant in ground that has been too well manured or dressed with fertilizers. Note also that cool, wet conditions may make petunias leafy, and check flowering. **5** Virus, various virus diseases (see Virus diseases). May attack established plants. **Symptoms:** mottling of leaves; spotting of leaves, with spotted areas dying; stunted plants and failure of flowers to open. **Control:** destroy badly affected plants. Note that some of the virus diseases of other plants, such as asters and tomatoes, may also affect petunias.

pH A logarithmic scale ranging from 1–14, measuring the acidity or alkalinity of a solution. A pH of less than 7 indicates that the solution is acidic (sour) and a pH of more than 7 indicates that the solution is alkaline (sweet), 7 being the neutral point. Going down the scale from 7, each digit represents a solution 10 times more acidic than the one before. Going up the scale from 7, each digit represents a solution 10 times more alkaline than the one before. The pH of a soil is actually the pH of the water in the soil, which contains soluble salts. Soils with a pH of between 5 and 7 are relatively fertile and provide congenial conditions for plants and the micro-organisms which are essential to the normal growth of plants. Plants have pH preferences. For example, rhododendrons and heathers prefer acid soil, with a pH of about 5. Brassicas need a soil which is neutral, pH 7, or slightly acidic, pH 6.5 to 7. To measure pH, various means are used. In laboratories, electric pH meters are used. Gardeners can check soil pH by using BDH indicator fluid, which reacts when a small quantity of soil is shaken up with water, giving a distinct colour change which can be checked against a colour chart. Kits of various sorts are available (see Soil testing). To raise pH, apply lime in one of its various forms (hydrated or ground limestone). Different types of soil will require differing amounts of lime to give a specific pH adjustment. To lower pH, apply flowers of sulphur to the soil, but again, soil type will determine the amounts required. Certain fertilizers, such as sulphate of ammonia, will tend to make soil more acidic.

Phlox Annual phlox varieties, derived from *Phlox drummondii*, border phlox varieties derived from *Phlox paniculata*, rock garden phlox varieties derived from *Phlox subulata*, and other phlox species. Hardy annuals and perennials. Height: creeping to medium. Flowering time: summer. Flower colours: cream, lilac, orange, pink, purple red or parti-coloured. Use: annual borders, herbaceous borders, low walls, rock gardens.

Pests and Diseases. 1 Caterpillars, Flax tortrix moth *Cnephasia interjectana*. May attack established plants in early summer. **Symptoms:** black or dark green caterpillars, which wriggle backwards when disturbed, feeding between top leaves, which the caterpillars tie together with silk. **Control:** spray with Trichlorphon. **2** Eelworm (see 6 below Stem eelworm). **3** Leaf spot, caused by the fungus *Septoria divaricata* (see Leaf spot). **4** Leafy gall (see Sweet pea 4 Leafy gall). **5** Powdery mildew, caused by the fungus *Erysiphe cichoracearum* (see Mildew). **6** Stem eelworm *Ditylenchus dipsaci* (see Eelworms). May attack at all stages. **Symptoms:** most leaves develop wavy edges and may be swollen while young leaves become thin and elongated; surface of the stems flakes away and large splits may appear. **Control:** destroy affected plants by burning them; obtain cuttings free of eelworm by washing roots thoroughly and taking *root* cuttings, which should be put into sterilized compost.

165

Physiological disorders   Physiological disorders are caused by
growing conditions either in the garden or in the greenhouse and
not by pests or bacterial, fungal or virus diseases. The main causes
of physiological disorders are: poor soil conditions such as excess
soil acidity or bad drainage; insufficient food supplies; unfavourable weather conditions; mechanical damage; incorrect use of
chemicals and inability of some plants to absorb certain elements;
and in the greenhouse, excess humidity, fluctuating temperatures,
inadequate ventilation, draughts, poor light, and irregular watering
and feeding. It is important to establish if a deficiency disease is
actually present and if so, to correct it in the right way. It is always
useful to ask the advice of the local horticultural adviser. Symptoms
of physiological disorders vary considerably and include failure to
develop and grow properly, stunted growth, leaf spotting, marking
and discolouration, failure to flower or poor flowers. These can
easily be confused with the symptoms of a pest or disease attack so it
is useful to be familiar with all common troubles. A further point is
that diseases such as root rot and stem rot may follow an initial
physiological disorder (see Root rot, Stem rot). When plants are
suffering from insufficient food supplies and are showing certain
characteristic symptoms, they are said to be suffering from
deficiency diseases or disorders, or mineral deficiencies. Different
plants require different minerals if they are to thrive and without
them, growth will be adversely affected. Here are some examples:
Tomatoes suffering from a lack of magnesium will show acute
bronzing of the leaves. Rhododendrons growing in chalky soil will
show acute leaf yellowing due to lime-induced iron deficiency, that
is, the lime in the soil prevents rhododendrons from taking up the
iron they require. The yellowing is invariably called lime-induced
chlorosis. In many cases, mineral deficiencies can be corrected
either by applying the element in short supply directly to the soil or
to the foliage of the plant as a foliar spray. However, it is difficult for
some plants to absorb certain elements. In such cases, chelated
elements (see Sequestrols) are useful in supplying the element in a
way that the plant can absorb without damage. The best way to
avoid physiological disorders is to know and understand the correct
growing conditions and requirements for specific plants.

Picea   (see Spruce)

Pine   *Pinus nigra* Austrian pine and varieties, *Pinus sylvestris* Scotch
fir or Scots pine and varieties and other *Pinus* species and varieties.

Mostly hardy, evergreen coniferous trees and shrubs. Height: trees small to large; shrubs prostrate and dwarf to medium. Cones: ornamental, varying in shape and size according to species. Foliage: decorative, varying in colour and form according to species. Variable and interesting habit of growth. Use: planted singly or in groups or avenues in large areas, windbreaks, smaller dwarf and prostrate sorts in borders, rock gardens.

Pests and Diseases. 1 Blister rust *Cronartium ribicola*. **Symptoms:** yellow blisters on branches and trunk in May/June; blisters burst, releasing spores, leaving scars and resin exudations; branches can in bad cases be killed. **Control:** if practical, spray with copper based fungicide. **2** Leaf cast *Lophodermium pinastri*. Attacks young trees. **Symptoms:** fungi attack fallen needles, causing infection of trees 1 to 3-years old. **Control:** spray young trees with copper based fungicide; give protection from cold winds, if practical, to reduce needle fall. **3** Needle rust, caused by various species of fungi. **Symptoms:** yellow-brown pustules on needles, causing a rusty appearance. **Control:** spray young trees with copper based fungicide. On older trees, the disease is seldom damaging. **4** Pine beetle *Myclophilus piniperda*. May attack trees of all ages. **Symptoms:** mature insects bore into bark, or into leading shoots; adults pupate in July and move to other trees. **Control:** keep ground free of old timber; in spring, spray with Malathion if practical. **5** Pine weevil *Hylobius abietis*. May attack young trees of all conifers. Most active in May, June and September. **Symptoms:** damaged bark, caused by weevils gnawing; eggs laid on dead wood where weevils hibernate in winter. **Control:** clear ground of dead timber; if on a small scale, spray ground with gamma-HCH during the hibernation period.

Plane *Platanus acerifolia* London plane and varieties, *Platanus orientalis* and varieties and other *Platanus* species. Hardy deciduous trees. Height: large. Flowering time: spring. Decorative fruit balls may hang until well into winter. Use: planted singly or in groups, avenues, London plane particularly useful for town and street planting, but also good near water.

Pests and Diseases. 1 Gloeosporium, caused by the fungus *Gloeosporium nervisequum*. May attack young and established plants. **Symptoms:** leaves marked with distinct patterns, with jagged edges and patches of fungus appearing on them; some defoliation may occur. **Control:** collect and burn fallen leaves;

spray young trees with Bordeaux mixture. **2** Leaf disease, caused by various species of fungi. May attack established plants. **Symptoms:** striking brown patterns develop on the leaves; light brown fungal pustules form on the dead areas. **Control:** collect and burn fallen leaves; spray young trees with copper-based sprays.

Plant breeding There has been, and always will be, considerable activity among plant hybridists and plant breeders endeavouring to produce plants with special qualities. In recent years the emphasis has been on the production of plants with inbred or natural pest and disease resistance, especially where it has been found difficult, impossible or undesirable to control the pests and diseases by chemical or other means. Research has been more evident in connection with crops of economic importance. There are many examples of successfully inbred resistance, notably, resistant rootstocks (see Resistant rootstocks) for tomatoes, tomato varieties resistant to certain strains of virus disease and to leafmould (a fungal disease), lettuce resistant to mildew, antirrhinums resistant to rust and many more.

Platanus (see Plane)

Plum Varieties of damson and plum derived from *Prunus domestica*. Hardy deciduous trees. Height: small to medium. Flowering time: spring. Flowers: white. Fruit: blue-black, red or yellow, varying in size and shape from small round to large oval. Use: food crop, grown for the fruit.

Pests and Diseases. 1 *Aphis*, (a) Damson and hop aphid *Phorodon humuli*, (b) Leaf curling plum aphid *Brachycaudus helichrysi*, (c) Mealy plum aphid *Hyalopterus pruni*. May attack at all stages. **Symptoms:** in all cases, aphids and honeydew present on the plants (see Honeydew), mottling on the leaves and growth is checked; the leaf curling plum aphid is the most spectacular, causing the leaves to curl and blacken. **Control:** spray with Dimethoate or Malathion at the white bud stage; where leaf curl occurs, use Dimethoate; use a tar oil wash in the dormant season. **2** Bacterial canker *Pseudomonas mors-prunorum* (see Bacterial diseases). May attack established plants. **Symptoms:** numerous spots with pale edges on the leaves, spots turning to small holes; branches and trunk show warped, split cankers which may extend the length of branches; gum oozes from the cankers, as it does in the case of cherry infected with bacterial canker, and gummosis may develop (see Cherry 12 Gummosis), but here the gum simply oozes out of the branches and trunk and there are no cankers. Plants may recover, but any cankers remain. **Control:** spray with Bordeaux mixture or a copper fungicide in spring and early summer to prevent infection; spray again in September and late October. **3** Blossom wilt, caused by the fungus *Sclerotinia laxa*. May attack established plants. **Symptoms:** blossom trusses wilt and become soft and brown; shoots may also wilt and die. May be troublesome in warm wet weather. **Control:** use a tar oil winter wash in the dormant season. **4** Brown rot, caused by the fungi *Sclerotinia fructigena, Sclerotinia laxa*. May attack fruit. **Symptoms:** fruits turn brown or black and look like prunes. **Control:** no control known, though winter washing helps; clear away all diseased fruit. **5** Brown scale insect (see 14 below Scale insects). **6** Caterpillars (see 11 below Plum fruit moth, 12 below Plum sawfly, 13 below Plum tortrix moth, 18 below Winter moth). **7** Fruit tree red spider mite *Panonychus ulmi*. May attack young and established plants. **Symptoms:** leaves silvered where mites feed; damage is similar to that to apples (see Apple 33 Fruit tree red spider mite); mites can be seen with a strong magnifying glass. **Control:** winter wash is the most effective method of control, but use Dimethoate, Malathion or any suitable insecticide in the growing season. **8** Leaf and bud mite (see Apple 38 Leaf and bud mite). **9** Leafhoppers. May attack young and established plants. **Symptoms:** mottling of leaves caused by greenfly-like insects attacking leaves and sucking sap. **Control:** spray in May/June at first stage of attack, using Dimethoate or Malathion; make sure sufficient spray is applied to cover the undersides of the leaves thoroughly; Nicotine and other chemicals may also be used. **10** Mussel scale (see 14 below Scale insects). **11** Plum fruit moth *Cydia funebrana*. **Symptoms:** red caterpillars make holes in the fruit and feed near the stone; brown liquid surrounded by black material called 'frass' oozes out of the holes; fruit is spoilt. **Control:** apply a tar oil wash thoroughly in winter; use Azinphosmethyl in spring. **12** Plum sawfly *Hoplocampa flava*. May attack fruit. **Symptoms:** caterpillars, creamy-white with brown heads, 12 mm ($\frac{1}{2}$ in) long, do great damage feeding on the fruit; holes made in the fruit are seldom visible. **Control:** spraying with Dimethoate or Fenitrothion at the cot split stage (pink bud opening) is adequate. **13** Plum tortrix moth *Hedya pruniana*. May attack young and established plants. **Symptoms:** the light green caterpillars tunnel into shoots from April till June; leaves are bound together with webs and caterpillars feed in this shelter. Habit of feeding makes control difficult. **Control:** spray mid-spring with Fenitrothion; use DNOC spring wash at the bud breaking stage; use tar oil wash in the dormant season. **14** Scale insects (see Scale insects), (a) Brown scale *Parthenolecanium corni*, (b) Mussel scale *Lepidosaphes ulmi*. May attack established plants. **Symptoms:** stems covered with small, hard, brown or reddish, inactive, shell-shaped insects. Scale insects are a sign of neglected trees. **Control:** spray with tar oil or DNOC plus Petroleum oil in the dormant season, or Malathion in the growing season. **15** Silver leaf, caused by the fungus *Stereum purpureum*. May attack young and established plants. **Symptoms:** leaves turn silvery, often one branch or one side of the tree being initially affected; later the branches die. This is a serious disease. The spores enter through any open wound. **Control:** prune in summer; paint over all cuts with tar oil wash or with paint; remove and burn dead branches (by April 1st in Scotland). **16** Thrips *Taeniothrips inconsequens*. May attack young and established plants. **Symptoms:** mottling of leaves due to thrips feeding; fruit may sometimes be attacked; very small insects are seldom seen. **Control:** spray with Dimethoate at the cot split stage (pink bud opening). **17** Wasps *Vespula* species. May attack fruit. Can cause considerable damage. **Symptoms:** wasps gathered round fruit and feeding on it; fruit hollowed out. **Control:** control nests with Destroyer (Carbaryl). **18** Winter moth *Operophtera brumata*. May attack established plants. **Symptoms:** (see Apple 53 Winter moth). **Control:** spray with Fenitrothion at the white bud stage; use tar oil winter wash in the dormant season.

Poinsettia *Euphorbia pulcherrima* varieties. Greenhouse deciduous shrub. Height: medium. Flowering time: throughout the year, according to method of growing, but chiefly at Christmas time. Flowers: insignificant, with large, showy, coloured bracts below them, which may be cream, pink or red. Use: greenhouse or indoor pot plant, commercial pot plant produced in bulk especially around Christmas time.

Pests and Diseases. 1 Black root rot, caused by the fungus *Thielaviopsis basicola*. **Symptoms:** rotting of roots, base of stems of cuttings and young plants. **Control:** use sterilized composts or soil-less media and clean containers (see Composts); after planting, give a Benomyl drench in place of the first watering. **2** Damping off, caused by various species of fungi. **Symptoms and Control** (see 1 above Black root rot). **3** Failure to produce coloured bracts (see Physiological disorders). **Symptoms:** bracts remain green. This trouble is caused by low temperature, combined with long hours of daylight and low light levels. **Control:** improve growing conditions by raising the temperature, keep the plants in good light, and follow commercial method of shading in summer to simulate a 12-hour day. **4** Leaf fall (see Physiological disorders). **Symptoms:** leaves drop off. **Control:** maintain a steady, suitable temperature; water carefully and regularly; provide suitable light.

Poison baits Products are available for the baiting of rats, mice, slugs and snails. One of the molluscicides, Methiocarb, will also suppress cutworms and woodlice. Baits should be used with great care because pets, wild birds and other gleaning animals may feed on the baits unless they are covered by slates in the case of molluscicides, and drainage tiles in the case of rodent killers. Using rolled oats or bran, it is possible to make up baits for the control of leatherjackets in turf.

Poplar *Populus alba* White poplar and varieties, *Populus balsamifera* Balsam poplar (aromatic), *Populus nigra* Black poplar and varieties including *Populus nigra italica* Lombardy poplar and *Populus tremula* Aspen (with trembling leaves). Hardy deciduous trees. Height: small to large. Flowering time: spring. Flowers: drooping catkins, often very attractive. Seeds: covered with cottony white down. Leaves: may be bright green, grey-green, pink-tinged, cream, golden-yellow or variegated and they colour to yellow in autumn; in some sorts the leaves have white undersides, in others they are aromatic, in others the leaf stalks are so slender the leaves quiver in the faintest current of air. Young growth in spring may be coppery. Habit of growth is variable; trees may be columnar, rounded, spreading or weeping. Use: very vigorous and quick-growing, useful in rows as screens or windbreaks, but they need ample space; good trees for town and coastal planting; Lombardy poplars and Balsam poplars make attractive avenues.

Pests and Diseases. 1 Aphis (see **7** below Root aphids). **2** Bacterial canker (see **4** below Canker and dieback). **3** Bark death and dieback, caused by the fungus *Dothichiza populea*. May attack established plants. **Symptoms:** dieback of branches; disease then affects the bark. **Control:** cut off branches; cut out affected areas of bark; paint all wounds with Arbrex. **4** Canker and dieback *Aplanobacter populi* (see Bacterial diseases). May attack young and established plants. **Symptoms:** branches dieback; cankers on branches and trunks. This is a serious disease of many species of poplar, especially where the growth of the trees is checked in any way by adverse weather or soil conditions. **Control:** cut out diseased wood and paint wounds with Arbrex or a similar chemical. **5** Leaf blister, caused by the fungus *Taphrina aurea*. May attack young and established plants. **Symptoms:** golden blister-like swellings, mainly on the undersides of the leaves. This disease is more unsightly than harmful. **Control:** if on a small scale, use copper-based sprays. **6** Leaf spots, caused by *Marssonina* species of fungi and by other species of fungi (see Leaf spot). **Symptoms:** black spots on leaves and shoots. Spotting is unsightly, and may be damaging to young trees. **Control:** it is impractical to spray large trees, but use copper-based sprays on young trees; regular pruning is helpful. Some varieties of poplar are more resistant than others. **7** Root aphids *Pemphigus* species. May attack young and established plants. **Symptoms:** attacks noticeable in spring; many of the root aphids found on other crops, for example lettuce, overwinter as eggs on poplars and construct reddish galls on the leaves and leaf stems in the spring, which disfigure the trees and form reservoirs of infection for summer vegetable crops. **Control:** chemical sprays are ineffective but if practical, hand-trim the galls, which are easily seen, to prevent disfigurement.

Poppy Annual poppy derived from *Papaver rhoeas* and other *Papaver* species, perennial poppy derived from *Papaver orientalis* and other *Papaver* species. Mostly hardy annuals and perennials. Height: small to medium. Flowering time: summer. Flower colours: cream, lilac, orange, pink, purple, red, white, yellow or parti-coloured. Use: annual borders, herbaceous borders, woodland gardens, rock gardens.

Pests and Diseases. 1 Downy mildew, caused by the fungus *Peronospora arborescens* (see Mildew). May attack at all stages. **Symptoms:** leaves covered with mealy growth. **Control:** improve general growing conditions.

Potato Varieties derived from *Solanum tuberosum*. Half-hardy perennial, with underground stem tubers. Height: medium. Tubers: vary in size and shape, may be large or small, round, oval or kidney shaped. Colours: creamy-white, blue-black, brown, red or yellow. Use: food crop, grown for the tubers.

Pests and Diseases. 1 Aphis, a number of different *Aphis* species, including the potato aphid *Macrosiphum euphorbiae* and the peach-potato aphid *Myzus persicae*. May attack young and established plants. **Symptoms:** aphids on the plants; distortion and wilting of shoots and leaves. Whether or not aphids cause serious damage depends on their numbers. In some areas, they are not regarded as a serious pest, whereas they may be in other areas. But one real danger is that certain species, especially the peach-potato aphid, may spread various virus diseases (see 50 below Virus diseases). **Control** (see Aphis). **2** Bacterial soft rot *Erwinia carotovora* (see Bacterial diseases). May affect stored tubers. **Symptoms:** rotting of stored tubers. Rot may follow on any damage to tubers. **Control:** lift tubers carefully and store under suitable conditions. **3** Black dot, caused by the fungus *Colletotrichum atramentarium*. May attack established plants. This disease may occur in association with eelworm attack (see 20 below Golden potato cyst eelworm, 31 below Pale potato cyst eelworm, 35 below Potato tuber eelworm). **Symptoms:** very similar to those of Silver scurf (see 40 below Silver scurf). **Control:** it causes little or no real damage, so can be ignored. **4** Blackleg *Erwinia atroseptica* (see Bacterial diseases). May attack young and established plants and stored tubers. **Symptoms:** stems may turn black and slimy about June at ground level; later tubers may also be affected, and may rot in store. This is a common trouble, but usually only a small number of plants are affected. **Control:** not of great consequence. **5** Black scurf and stem canker, caused by the fungus *Rhizoctonia solani*. May attack established plants and tubers. **Symptoms:** rolling of the upper leaves of plants; in July, a band of white fungal growth round stems at ground level but fungus does not penetrate into the stems; at lifting time, black resting bodies of the fungus on the skins of tubers. Not really a disease of any great consequence unless infected potato tubers are planted. **Control:** soak tubers wanted for seed in organo-mercury compound for 15 minutes to kill spores. **6** Blight, caused by the fungus *Phytophthora infestans*. May attack established plants and tubers. A serious disease. **Symptoms:** dark spots on the tips and margins of the leaves, spreading over the leaves; whitish mould on the undersides of the leaves; premature destruction of leaves; rusty marking on skins of stored tubers, and deteriorating flesh. This disease is more common in wet than in dry areas and may be especially bad in wet summers, when premature loss of leaves may lead to reduction of total yield, and spores present on the tubers may cause rotting in store. The disease does not pass from tuber to tuber in store, but secondary diseases may cause rotting which may spread. Tubers which appear to be fit for planting may be carrying resting spores of the fungus, and, after planting, these may give rise to infection. Blight spreads according to weather conditions, and commercial growers are warned of blight risk so that they can use preventive sprays. **Control:** avoid planting doubtful tubers and spray in good time, by reference to local farmers, probably in June/July; use some of the many chemicals available, the most suitable of which for amateurs are copper compounds, Maneb or Zineb; destruction of the haulms (potato plant tops) by chemical means some time before lifting is also useful in preventing fungal spores from getting down to the tubers; blight resistance of certain varieties is a possibility. The *Ministry of Agriculture Agricultural Chemicals Approval Scheme* booklet lists those chemicals available to the commercial grower. **7** Capsid bugs. May attack young and established plants, causing a shot-hole appearance of the leaves where they feed (see Capsid bugs). **8** Coiled sprout (see Physiological disorders). May attack young plants. **Symptoms:** roots develop into a tight ball, checking growth. This condition is attributed to cold weather after planting. **Control:** nothing can be done beyond having the ground in the best possible condition, and not planting too early for the particular area. **9** Colorado beetle *Leptinotarsa decemlineata*. May attack established plants. This pest is so well publicized that it scarcely needs description. **Symptoms:** beetles, 9 mm ($\frac{3}{8}$ in) long, with black and yellow *stripes*, not to be confused with ladybirds, which are *spotted*. The females lay eggs on leaves and these hatch out into dark red larvae with humped backs which, if present in sufficient numbers, are very destructive indeed. Plants can be completely stripped of leaves, stopping all growth. The adult beetles hibernate in the soil in winter at a depth of 25–30 cm (10–12 in). A native of North America, the pest has bred in Europe for some years, but fortunately not in large numbers. This is a notifiable pest, which means contacting the police and in turn the Ministry of Agriculture, who will destroy the beetles and the crops of potatoes involved, and protect neighbouring crops of potatoes. **Control:** organo-mercury compounds used by the appropriate authorities. **10** Common scab, caused by the fungus *Streptomyces scabies*. May affect the tubers. **Symptoms:** ragged scabs on the surface of the tubers. The scabs are unsightly but superficial. **Control:** avoid liming ground for potatoes; use varieties which show some resistance, such as Arran Pilot, Golden Wonder, Maris Peer and Pentland Crown. **11** Corky scab (see 36 below Powdery scab). **12** Cutworms (see Cutworms). **13** Deficiency disorders (see Mineral deficiencies). May occur in young and established plants. Potatoes, like tomatoes, are very susceptible to deficiencies of major and micro elements, and affected plants show typical symptoms. For instance: with magnesium deficiency, leaves dying back; with manganese deficiency which is common on acid soils, yellowing and stunting of the plants; with potash deficiency, marginal scorching of the leaves. **Control:** in severe cases, soil analysis may be necessary, and spraying with suitable chemicals is a possibility, but it is best to ask the advice of the local horticultural adviser. **14** Dry rot, caused by *Fusarium* species of fungi especially *Fusarium caeruleum*. May affect stored tubers. **Symptoms:** shrivelling and rotting of stored tubers which may show wrinkled patches with pale blue, pale pink or white fungal spores. **Control:** take great care at lifting time to avoid damage to the tubers; plant sound seed potatoes only. The chemical Tecnazene, put on as directed *before* tubers are stored, will both check too early sprout growth in store, and control dry rot. **15** Early blight, caused by the fungus *Alternaria solani*. May affect established plants. **Symptoms:** ringed spots on the leaves. May occur in hot dry summers. **Control:** does not cause any real damage in Britain. **16** Eelworm (see 20 below Golden potato cyst eelworm, 31 below Pale potato cyst eelworm, 35 below Potato tuber eelworm). **17** Flea beetles (see Flea beetles). **18** Frost damage (see Physiological disorders). May affect young plants. **Symptoms:** blackening or softening of the leaf margins. **Control:** little can be done beyond planting at dates suitable for particular districts. **19** Gangrene, caused by the fungi *Phoma foveata*, *Phoma solanicola* and other *Phoma* species. May affect stored tubers. **Symptoms:** very similar to those of dry rot (see 14 above Dry rot) but fungal patches are smooth. Low temperatures and dryness favour the development of this disease. **Control:** take great care at lifting to avoid damage to the tubers; plant sound seed potatoes only. **20** Golden potato cyst eelworm, Potato root eelworm, Potato cyst eelworm *Globodera (Heterodera) rostochiensis* (see Eelworm). May attack young and established plants. A serious and persistent pest which can cause considerable loss. **Symptoms:** vary with the level of infestation; in a really severe attack the foliage yellows and dies back and the tubers, when lifted, may be the size of marbles. To confirm an attack, lift a plant carefully and look for brown, white or yellow pinhead sized cysts adhering in quantity to the roots (see Eelworms). The eelworm larvae cannot be seen with the naked eye, and should not be confused with various visible pests. **Control:** crop rotation is the best control, except perhaps for early potatoes (see Rotation); of the chemicals recommended, such as Aldicarb, Basamid, Dichloropropene and mixtures, and Oxamyl, only Basamid is likely to be practical and safe for the amateur. When the levels of eelworm are low, the use of plenty of organic matter and gross feeding can enable potatoes to be grown reasonably well, despite the presence of eelworm. There is some resistance in certain varieties such as Maris Piper and Pentland Javelin, but the situation is a changing one, and gardeners are advised to find out the up-to-date situation when ordering potato seed. **21** Grey mould, caused by the fungus *Botrytis cinerea* (see Botrytis). May attack established plants. **Symptoms:** rotting of potato tops; grey mould visible. This disease may occur in wet summers. **Control:** seldom serious. **22** Growth mottle (see 50b below Virus diseases). **23** Herbicide damage (see Physiological disorders). May occur on young and established plants. **Symptoms:** abnormal growth, may be due to drift of hormone type weedkillers (see Hormone). Chemical weedkillers, especially paraquat applied post-emergence, and residual weedkillers, can give rise to mottling of the leaves. **Control:** use weedkillers, but follow all directions carefully in suitable weather conditions. **24** Lack of growth. **Symptoms:** failure of tubers to sprout after planting. This may be due to various diseases affecting the tubers. **Control:** it is very important to obtain good, clean, reliable potato seed and to examine it carefully *before* planting. Doubtful seed should *not* be planted. **25** Leaf curl (see 50c below Virus diseases). **26** Leaf roll (see 50c below Virus diseases). **27** Little potato (see Physiological disorders). **Symptoms:** tubers, instead of making normal growth, remain small, and cluster together at the base of the foliage. This condition is largely due to cold weather after planting. **Control:** try to plant at dates suitable for particular districts. **28** Millipedes (see Millipedes). May enlarge holes made by slugs (see Slugs and snails) but are generally not capable of actual attack on healthy tubers. **29** Mosaic (see 50a, 50d below Virus diseases). **30**

Mottle (see 50b, 50f below Virus diseases). **31** Pale potato cyst eelworm (see 20 above Golden potato cyst eelworm). **32** Pink rot caused by the fungus *Phytophthora erythroseptica*. Affects tubers in store. **Symptoms:** pink rotting at the heel end of stored tubers. **Control:** practise long rotation (see Rotation). **33** Potato cyst eelworm (see 20 above Golden potato cyst eelworm). **34** Potato root eelworm (see 20 above Golden potato cyst eelworm). **35** Potato tuber eelworm *Ditylenchus destructor*. May attack tubers. **Symptoms:** discoloured patches on the tubers, similar to those caused by blight (see 6 above Blight). This eelworm is harboured by two main hosts, field mint and corn sowthistle (see Host). **Control:** it is unlikely to be of importance, but if these weeds are in evidence, they should be destroyed. **36** Powdery scab or Corky scab, caused by the fungus *Spongospora subterranea*. May affect tubers. **Symptoms:** scales containing snuff-like powder, the spore balls of the fungus, on the surface of the tubers; warts not unlike those of wart disease caused by spores (see 51 below Wart disease), but powdery scab is not a serious disease. **Control:** do not plant affected tubers; avoid liming ground for potatoes, as this disease is worse in limy soils. **37** Rosy rustic moth *Hydroecia micacea*. May attack young and established plants. **Symptoms:** stems collapse and show tunnelling at ground level, with pinkish caterpillars in the tunnels in June and July. The small moths have banded pinkish wings. Eggs are laid in late summer and the caterpillars overwinter in the soil and start to attack the following summer. The caterpillars feed on various weeds, such as couch grass, docks, horsetails and plantains, but may also attack cultivated plants such as potatoes and tomatoes. **Control:** collect and destroy tunnelled stems; remove weeds. **38** Scab (see 10 above Common scab, 36 above Powdery scab). **39** Second growth (see Physiological disorders). **Symptoms:** knob-like growths develop from the tubers or tubers may be oddly distorted. Alternate spells of wet and dry weather, or wet weather following a very dry spell may cause this type of growth. **Control:** ensure that ground has ample organic matter, which will help to minimize the effect on growth of wet and dry weather. **40** Silver scurf, caused by the fungus *Helminthosporium atrovirens*. May affect stored tubers. **Symptoms:** small silvery patches on the tubers which increase in size during storage and cause skin to peel. **Control:** not a serious disease and should not cause undue trouble, but avoid planting affected tubers. **41** Skin spot, caused by the fungus *Oospora pustulans*. May affect stored tubers. **Symptoms:** pimple-like growths over the surface of tubers, and although they are superficial and do not affect tubers for eating, the eyes may be injured and may fail to sprout. There has been considerable research on the effect and control of this disease, and it has been established that certain varieties are very susceptible while others, such as Golden Wonder, are highly resistant. **Control:** little can be done about controlling this disease, apart from following the practice of sprouting seed potatoes and discarding any doubtful ones. This trouble should always be reported to the supplier. **42** Slugs (see Slugs and snails). Note that slugs may do serious damage and may ruin the whole crop, especially in wet summers and where organic matter has been used freely near the surface of the soil. **Control:** although baits of bran and metaldehyde, and pellets based on metaldehyde or Methiocarb are frequently recommended (see Poison baits), early lifting, especially in wet seasons, is probably the best control. **43** Spraing (see 50e below Virus diseases). **44** Stalk break, caused by the fungus *Sclerotinia sclerotiorum*. May attack established plants. **Symptoms:** breaking over of stalks at soil level. This is a disease almost entirely confined to West Ireland. **45** Stem canker (see 5 above Black scurf and stem canker). **46** Stem mottle (see 50f below Virus diseases). **47** Veinal necrosis (see 50g below Virus diseases). **48** Verticillium wilt, caused by the fungus *Verticillium albo-atrum*. May attack established plants. Is soil and tuber borne. **Symptoms:** wilting, caused by the fungus invading the conducting tissue of the stems. This disease is rarely reported on potatoes. **Control:** (see Tomato 44 Wilts); practise crop rotation rigorously; obtain new seed for the next season. **49** Violet root rot, caused by the fungus *Helicobasidium purpureum*. May affect tubers. **Symptoms:** net-like brown or purple strands of the fungus form on the tubers. This disease attacks a wide range of crops, including carrots (see Carrot 12 Violet root rot), but is not a serious trouble of potatoes. **Control:** discard badly affected tubers; do not use affected tubers as seed potatoes. **50** Virus diseases (see Virus diseases). A number of virus diseases affect potatoes, and serious attacks can reduce yields to virtually nil. The presence of virus diseases in planting stock is one reason why the Certification Scheme for seed potatoes was implemented (see Certified stock) and it is because of these diseases that it is advisable not to save home seed for too long, but to buy in new seed potatoes from a reliable source. The following are some of the main virus diseases, with varying symptoms present on the leaves, in the size of the plants or on the tubers. Note that the virus diseases listed below are, with the exception of one form of spraing, tuber-borne, but symptoms do not become *evident* until the potatoes are producing foliage. (*a*) Aucuba mosaic. **Symptoms:** yellow mottling on the leaves, similar to the mottling on the leaves of the variegated form of the shrub *Aucuba japonica*. (*b*) Growth mottle. **Symptoms:** yellow blotching of the leaves. (*c*) Leaf roll or leaf curl. Possibly the most common and most serious of the virus diseases. **Symptoms:** leaves roll upwards and the lower leaves are brittle. Leaf roll is spread by aphids. Some varieties are more susceptible than others. In Golden Wonder, for instance, symptoms are severe. Do not confuse the symptoms of leaf roll in potatoes with the leaf rolling or curling seen in tomatoes, which is caused by temperature variations (see Tomato 25 Leaf curling). (*d*) Mosaic. Severe and mild Mosaic. **Symptoms:** varying degrees of leaf mottling, spotting and streaking, depending on which strain of virus is present (see Strain and 50a above). (*e*) Spraing. **Symptoms:** rings of brown dead tissue inside the tubers. (*f*) Stem mottle. **Symptoms:** dwarfing and stunting of plants. (*g*) Veinal necrosis. **Symptoms:** foliage paler and more open than normal. (*h*) Virus E, Viruses S and M. **Symptoms:** crinkling of the leaves. (*i*) Witches' brooms. **Symptoms:** congested growth said to resemble brooms. **Control:** little can be done on a garden scale to control various virus diseases beyond buying clean seed potatoes from a reliable source, controlling aphids and other insects known to spread virus diseases and practising rotation (see Rotation). **51** Wart disease, caused by the fungus *Synchytrium endobioticum*. May attack the tubers. Formerly a very serious disease, but immune varieties now available. **Symptoms:** large warts on the tubers, which become distorted beyond recognition, and often end by rotting away. **Control:** there is no known chemical control, but a great many varieties are resistant to this disease and only those should be planted if there is any risk of disease attack. For instance, Epicure and King Edward are *not* immune. Note that *all* new varieties are immune, *eg* Maris Piper and Pentland Javelin. **52** Watery wound rot, caused by the fungus *Pythium ultimum*. May affect the tubers. **Symptoms:** tubers rot and exude a thin watery liquid, until only papery skins of the tubers remain. This disease is only likely to affect tubers which are lifted too early and are harvested under hot, dry conditions. **Control:** follow correct lifting or harvesting procedures. **53** Wireworms (see Wireworms). **Symptoms:** tubers holed. **Control:** it is unlikely that wireworms will cause great loss, especially if early lifting is practised; use Bromophos, except on newly-worked grassland. For commercial growers, Aldrin and Phorate are recommended. **54** Witches' brooms (see 50i above Virus diseases). **55** Yellow spotting (see Physiological disorders). **Symptoms:** yellow spotting and mottling on the leaves. This condition may be due to periods of adverse weather, for example, cold nights and warm days, but it is probable that a virus disease may also be present (see 50 above Virus diseases). Chemical weedkillers may also cause mottling of the leaves (see 23 above Herbicide damage).

Primula   There is a wide range of hardy primulas, such as *Primula denticulata*, *Primula japonica*, *Primula rosea* and the primrose *Primula vulgaris*, with varieties of all these species. *Auricula* varieties are derived from *Primula* species. Polyanthus varieties are derived from a cross between the primrose and the cowslip *Primula veris*. Glasshouse primulas include varieties of *Primula malacoides*, *Primula obconica* and *Primula sinensis*. Hardy herbaceous perennials. Height: small. Flowering time: varies according to method of growing and sort throughout the year. Flower colours: cream, blue, lilac, pink, purple, white, yellow or parti-coloured. Use: beds, borders, rock gardens, water gardens, woodland gardens, spring bedding, outside containers, window boxes, commercial cut flowers, greenhouse and indoor pot plants.

Pests and Diseases. 1 Aphis, Auricula root aphis *Pemphigus auriculae*. May attack at all stages. Auricula varieties mainly affected. **Symptoms:** patches of white dust on the roots; aphids, which are bloated and greyish in colour, also on the roots; poor growth and wilting of plants in bright sunlight. **Control:** re-pot potted plants in clean compost; drench soil in borders with Diazinon. Systemic aphid sprays do not work against root aphids (see Systemic). **2** Bryobia mite *Bryobia* species. May attack at all stages. **Symptoms:** large number of green, globular mites with hairy bodies on plants. **Control:** spray with Dicofol or Malathion. Commercially, Demeton-S-methyl may be used. **3** Cutworms (see Cutworms). **4** Downy mildew, caused by a *Peronospora* species of fungus (see Mildew). May attack young and established plants.

Symptoms: spots on leaves, pale on upper surfaces, and with grey mould on lower surfaces. **Control:** remove and burn all affected leaves; dust plants with a copper lime fungicide. **5** Glasshouse whitefly *Trialeurodes vaporariorum* (see Whitefly). *Note :* primulas must not be sprayed under direct sunlight, or when compost is very dry, or the foliage will be scorched. **6** Grey mould, caused by the fungus *Botrytis cinerea* (see Botrytis). May attack at all stages. **Symptoms:** rotted leaves. **Control** (see Botrytis). **7** Leaf spot, caused by fungi including *Cercosporella primulae* and *Ramulariae primulae*. May attack established plants. **Symptoms:** brown spots on leaves. Not serious. **Control:** if troublesome, remove spotted leaves; improve growing conditions. **8** Slugs (see Slugs and snails). **9** Virus, Mosaic virus, Tomato spotted wilt virus (see Virus diseases). **Symptoms:** scorching and yellowing of the leaves; stunting; failure to flower. Affects glasshouse primulas chiefly. **Control:** destroy all infected plants. **10** Whitefly (see 5 above Glasshouse whitefly).

Privet *Ligustrum ovalifolium* and varieties, *Ligustrum vulgare* and varieties, and other *Ligustrum* species and varieties. Hardy deciduous and evergreen shrubs and trees. Height: trees, small; shrubs, medium. Flowering time: summer. Flower colours: cream or white. Flowers: fragrant. Berries: black or blue-black. Foliage: variegated. Use: borders, hedges, street planting.

Pests and Diseases. 1 Dieback. **Symptoms:** gradual death of a section of privet hedge. There can be many causes, all of which should be investigated. (a) Gas, town or natural, the latter asphyxiating the plants. **Control:** check with the local branch of the Gas Board to see if there is any escape of gas. (b) Toxic chemicals, including weedkillers or products used for pavements or roads. **Control:** check on use of any such materials. (c) Diseases caused by various fungi, including *Armillaria* (see Armillaria mellea) which is easy to identify and others such as *Phytophthora* species and *Verticillium* species, which can require the help of a plant pathologist to diagnose. **Control:** check possible diseases, if necessary with the help of a plant pathologist, through your local horticultural adviser. (d) Aphis infestation, which is a likely cause of severe leaf fall. **Control:** if aphids are present, spray with a suitable insecticide (see Aphis). In general, remove all dead branches or dead bushes and if replanting is necessary, renew soil locally before doing this. **2** Lilac leaf miner *Caloptilia syringella*. May attack established plants. **Symptoms:** blisters caused by larvae tunnelling within leaves; growth is checked. **Control:** spray with gamma-HCH, Malathion or Nicotine in spring. **3** Privet thrips *Dendothrips ornatum*. May attack established plants. **Symptoms:** severe speckling of the leaves with distortion of new growth, becoming noticeable about midsummer. **Control:** trim hedge thoroughly to expose inner leaves and then spray with gamma-HCH. Note that this pest is confined to southern England.

Prunus The genus *Prunus* (see Genus) includes a wide range of plants, some ornamental, some grown as food crops. As happens with many garden varieties of plants, especially those used as food crops, the actual species are not grown, but varieties derived from them are (see Variety). The following list includes some of the more familiar plants in the genus *Prunus*: *Prunus armeniaca*, parent of the apricot; *Prunus avium*, the gean or wild cherry, a parent of the sweet cherry; *Prunus cerasifera*, the Myrobalan or cherry plum; *Prunus cerasus*, a parent of the Morello cherry; *Prunus domestica*, parent of the plum; *Prunus dulcis*, the common almond; *Prunus laurocerasus*, the common laurel or cherry laurel; *Prunus lusitanica*, the Portugal laurel; *Prunus padus*, the bird cherry; *Prunus persica*, a parent of the peach; *Prunus serrulata*, a parent of the Japanese cherry. Deciduous or evergreen, trees or shrubs. Use: varieties grown as food crops, for their ornamental value in flower, fruit or foliage.

For **Pests and Diseases** (see Plum).

Pyracantha *Pyracantha coccinea* Firethorn, and other *Pyracantha* species and varieties. Hardy evergreen shrubs. Height: large. Flowering time: spring. Flower colour: white. Berries: orange-red or red. Stems: spiny. Use: planted singly or in groups, often against walls, pillars and fences.

Pests and Diseases. 1 Scab, caused by the fungus *Fusicladium pyracanthae*. May attack established plants. **Symptoms:** berries covered with a dark brown coating of fungus, which looks similar to scab on apples and pears. **Control:** spray with lime sulphur in March and April.

R

Rabbit *Cuniculus lagopus* is a serious and very destructive pest. Rabbits feed on a wide range of plants, both soft-leaved and woody. They cause considerable damage to trees by gnawing the bark.

Control: if possible, put up suitable wire fencing to protect plants; put boards connected by hooks and eyes around the stem or trunk of individual trees; alternatively, put sticks 2.5 cm (1 in) in diameter round trees; for young trees, use proprietary rabbit guards; other measures include dusting plants with garden pepper, shooting and gassing; spray the lower part of trees with the repellent Anthraquinone.

Radish Varieties derived from *Raphanus sativus*. Hardy annual, with fleshy root. Height: small. Colours: red, white, yellowish or parti-coloured. Use: food crop, grown for the roots.

For **Pests and Diseases**, see Brassicas.

Rainfall This varies according to latitude, altitude, region, and local placement in relation to airstreams and hill formation. In Britain, rainfall is generally higher in the west and lower in the east, the region of lowest rainfall being the south east of England. Incidence as well as total rainfall is a critical factor in relation to plant growth and crop production. Soil type and the ability of soil to conserve rainfall, as well as natural water tables, which are influenced by rivers and large expanses of water, are also important. Sunshine intensity and duration, prevailing temperatures and exposure to wind must be taken into account when considering the climatic factors which affect the evaporation of rainfall. The combination of rainfall and temperature, coupled with soil type, causes crops and plants to perform well in some regions and fail in others. Crop timing is also influenced greatly by rainfall.

Raspberry Varieties derived from *Rubus idaeus* Wild raspberry. Hardy deciduous shrubs. Height: medium. Flowering time: spring. Fruit: red or yellow. Stems: usually prickly. Use: food crop grown for the fruit, trained to supports.

Pests and Diseases. 1 Aphis, Raspberry aphid *Aphis idaei*, Rubus aphid *Amphorophora rubi* (see Aphis). May attack at all stages.

183

Symptoms: large, long-legged, pale green aphids feed on the undersides of the leaves, which turn mottled and, in the case of the raspberry aphid, curled; later in the season the raspberry aphid turns creamy white. Certain aphids may spread virus diseases (see 20 below Virus) which are more serious than the damage done by the aphids themselves. **Control:** spray as required during the summer with Derris, Dichlorvos, Dimethoate, Formothion or Malathion; spray with tar oil or DNOC wash in the dormant season. **2** Blackberry mite *Acalitus essigi* (called Redberry disease in USA). May attack established plants. **Symptoms:** some of the drupelets (the individual parts which make up fruits such as blackberries or raspberries) fail to change from red to deep purple when ripe. This pest is difficult to control completely. **Control:** remove affected canes after fruiting. The spray recommended commercially is Endrin, which is not available to amateur gardeners. **3** Blossom weevil *Anthonomus rubi*. May attack

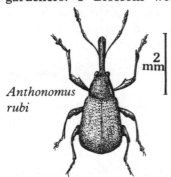

Anthonomus rubi

2 mm

established plants. **Symptoms:** weevils, similar to apple blossom weevil (see Apple 2 Apple blossom weevil) but without the V marking on the back, pierce unopened blossoms or tips of flower stalks and lay eggs; the larvae feed in the buds and later the weevils bite pieces out of the leaves. **Control:** spray with Carbaryl when the damage is first seen. **4** Botrytis, Grey mould, caused by the fungus *Botrytis cinerea*. May attack established plants. **Symptoms:** fruit is reduced to a rotten black mass covered with grey mould. The damage to fruit can be serious, especially in wet seasons. **Control:** spray regularly starting at 10% flowering (the beginning of flowering), using Benomyl, Dichlofluanid or Thionate-methyl. **5** Bramble shoot webber *Notocelia uddmanniana*. May attack young and established plants. Adult moth has pale brown green-tipped wings. **Symptoms:** sluggish dull brown caterpillars bind together the leaves and tips of canes in April with webbing, causing the death of terminal buds. **Control:** spray with Carbaryl or Azinphos-methyl in mid-April. **6** Cane blight, caused by the fungi *Fusarium avenaceum, Leptosphaeria coniothyrium* and other species of fungi. May attack established plants. **Symptoms:**

sudden dieback of canes, due to fungal attack at the bases. The disease gains entry through cracks caused by insects, such as Raspberry cane midge (see 15 below Raspberry cane midge), or by cold weather. Insecurely tied canes may be rocked in high winds and loosened in the ground, which allows water to lie at the bases. This encourages disease attacks. **Control:** as soon as damage is seen, cut out and burn diseased canes; secure canes properly, and do not allow raspberry cane midge, or diseases such as cane spot (see 7 below Cane spot) to make headway. **7** Cane spot, caused by the fungus *Elsinoe veneta*. May attack established plants. **Symptoms:** purplish spots on young canes in early summer enlarge, turn pale and sunken, and stems crack; leaves may be spotted and fruit misshapen; development of buds may be affected. **Control:** spray at bud burst when shoots are not more than 12 mm ($\frac{1}{2}$ in) long, again before leaves have unfolded and yet again before blossom, using Benomyl, a copper fungicide, Dichlofluanid, lime sulphur or Thiram. Control is similar to that for spur blight (see 18 below Spur blight). **8** Clay-coloured weevil *Otiorhynchus singularis* (see Blackcurrant 9 Clay-coloured weevil). **9** Grey mould (see 4 above Botrytis). **10** Lateral wilt, caused by the fungus *Fusarium avenaceum*. May attack established plants. **Symptoms:** laterals fail to grow in spring. **Control:** none known, but cut out and burn affected canes. **11** Leafhoppers (see 17 below Rubus Leafhoppers). **12** Purple blotch, caused by the fungus *Septocyta ramealis*. May attack established plants. **Symptoms:** purple blotches on raspberry canes coalesce, causing failure of laterals and general poor growth. **Control:** cut out and burn seriously affected canes; spray with a copper fungicide. **13** Raspberry beetle *Byturus tomentosum*. May attack established plants. **Symptoms:** beetles first appear in May, browny-yellow in colour, later becoming grey-brown, and 4 mm ($\frac{1}{6}$ in) in length; beetles destroy many of the blossom buds, and later feed on the fruit, causing misshapen berries; eggs are laid and in June, larvae attack fruitlets, tunnelling in the flesh and plugs; when mature, the larvae overwinter in the soil as chrysalids. This is a common pest of blackberries, loganberries and raspberries. **Control:** spray in early April with Azinphos-methyl commercially, Carbaryl, Derris, Fenitrothion or Malathion. **14** Raspberry borer, Raspberry moth *Lampronia rubiella*. May attack young and established plants. **Symptoms:** dead cane tips, if examined at the end of April or early May, may contain red larvae or brown chrysalids; canes die due to buds being eaten away. This is a serious pest. **Control:** spray with tar oil or DNOC winter wash in the dormant season; in April spray with Carbaryl. **15** Raspberry cane midge *Resseliella theobaldi*. May attack established plants. **Symptoms:** pink legless larvae feed on the surface of the canes, often causing them to split, which allows diseases to enter. **Control:** spray with gamma-HCH, delivering the spray to the lower parts of the canes to avoid fruit taint. **16** Redberry (see 2 above Blackberry mite). **17** Rubus leafhoppers *Macropsis fuscula, Macropsis scotti*. **Symptoms:** active green insects feeding on the berries. **Control:** spray with Dimethoate or Malathion as soon as the pests are seen. **18** Spur blight, caused by the fungus *Didymella applanata*. May attack established plants. **Symptoms:** failure of fruiting buds, especially in wet seasons, with overcrowded canes or with canes which are too soft owing to excess nitrogen feeding. **Control:** cut out and burn seriously affected canes; spray with Benomyl or Dichlofluanid when the buds are not more than 12 mm ($\frac{1}{2}$ in) in length, and repeat at 14-day intervals until the blossom period is past. **19** Stamen blight, caused by the fungus *Hapalosphaeria deformans*. May attack established plants. **Symptoms:** flowers seem to be more open and flattened than normal; fruit development uneven. **Control:** spray with Captan or Dichlofluanid 14 days and again 21 days after the start of flowering; do not use Captan if raspberries are for canning or preserving. **20** Virus (see Virus diseases), various diseases, for example, Bushy dwarf and Severe stunt in raspberry, which cause severe stunting, Raspberry mosaic, where the leaves show light and dark mottling, spotting or vein banding followed by stunting and poor growth, and Yellow disease, where leaves and fruit are a golden yellow. **Control:** apart from controlling various sucking pests, such as aphids, which may spread virus diseases, little can be done; remove and burn badly affected plants; buy certified canes (see Certified stock) and plant them in good ground which has not previously grown cane fruit. **21** Weevil (see 3 above Blossom weevil, 8 above Clay-coloured weevil).

Red pepper   (see Pepper)

Red spider mite   (see Mites)

Red and white currants   Varieties derived from hybridizing 3 *Ribes* species, possibly others also (see Hybrid). Hardy deciduous shrubs. Height: medium. Flowering time: spring. Fruit: red or white. Use: food crop, grown for the fruit.

For **Pests and Diseases**, see Blackcurrant.

Resistant rootstocks Many plants are grafted or budded on to rootstocks (see Stocks) which have inbred resistance to certain pests and diseases. This means that scions (the parts put on to the rootstocks) of desirable varieties may acquire roots resistant to certain pests and diseases. In addition, the rootstocks can also confer extra vigour on the scion varieties. Examples are grafting of tomato or cucumber varieties on to rootstocks resistant to eelworms, and apple varieties on to rootstocks resistant to woolly aphid attack.

Resistant varieties   (see Plant breeding)

Rhododendron   *Rhododendron* and *Azalea* species and varieties. The genus *Rhododendron* now includes *Azalea*. It is a very large genus and many of the species are grown. There are also a very large number of hybrids (see Hybrid). Hardy, not fully hardy and greenhouse, evergreen and deciduous trees and shrubs. Height: shrubs, prostrate to large; trees, small to medium. Flowering time: early spring to early summer in the open, varying in the greenhouse. Flower colours: blue, cream, orange, pink, purple, red, white, yellow or parti-coloured. Leaves: may be decorative and in some species, very large. Use: planted singly or in groups, borders, informal hedges, windbreaks, naturalized in woodland, rock gardens, greenhouse pots, borders, forced as commercial glasshouse pot plants especially for Christmas.

Pests and Diseases. 1 Azalea leaf miner *Caloptilia azaleae*. May attack indoor azaleas and those planted outside in mild climates at all stages. **Symptoms:** blister-like mines in the leaves caused by caterpillars; premature falling of leaves. **Control:** spray with gamma-HCH as soon as leaf damage is seen and twice at fortnightly intervals thereafter. **2** Azalea whitefly *Pealius azaleae* (see Whitefly). May attack evergreen azaleas at all stages. **Symptoms:** small white-winged insects feeding on the undersides of leaves, along with their scale-like young stages; honeydew and sooty moulds on the lower leaves (see Honeydew). **Control:** drench frequently with water to remove the honeydew; pesticide is rarely necessary. **3** Bud blast (see 19 below Rhododendron bud blast). **4** Bud death (see 16 below Leaf scorch and dead buds). **5** Bud drop (see Physiological disorders). May occur on established plants. **Symptoms:** buds drop without opening. This is mainly a trouble of azaleas forced in commercial glasshouses, though it may also occur when such plants are brought indoors. **Control:** avoid draughts and marked changes in temperature; water plants freely, but do not over-water to the point of continual saturation. **6** Chlorosis (see Physiological disorders) caused by iron deficiency. May occur on young and established plants. **Symptoms:** yellowing between the veins, especially on younger leaves. **Control:** water with chelated iron solution at the recommended rate (see Sequestrols); mulch heavily with peat to reduce pH, or in pots, use suitable compost; apply sulphur to the ground in extreme cases to reduce soil pH. **7** Clay-coloured weevil *Otiorhynchus singularis*. May attack outdoor azaleas and rhododendrons at all stages. **Symptoms:** adult weevils 6–10 mm ($\frac{1}{3}$ in) long and speckly brown in colour, gnawing at the bark of young shoots, causing them to wither; large holes chewed in the leaves and blossom in early summer. **Control:** shake the branches vigorously and collect the weevils in an inverted umbrella; moderate control is achieved by dusts or sprays of gamma-HCH when fresh damage is seen. **8** Gall, caused by the fungus *Exobasidium vaccinii*. May attack established plants. **Symptoms:** small, hard reddish swellings on leaves and flower buds, which later become covered with a white bloom of fungal spores. This is one of the best-known and most spectacular diseases of azaleas and rhododendrons. **Control:** remove and burn the affected parts as soon as the disease is seen, and before the white bloom of fungal spores can develop; spray with Bordeaux mixture or other copper fungicide. **9** Glasshouse thrips *Heliothrips haemorrhoidalis*. May attack at all stages. **Symptoms:** small brown adults and yellow larvae on the plants; white patches and streaks on the leaves where thrips have been feeding. **Control:** use gamma-HCH or Malathion smokes, two applications at 14-day intervals. **10** Glasshouse whitefly *Trialeurodes vaporariorum* (see Whitefly). May attack at all stages. **Symptoms:** masses of small white winged adults on the plants, rising in clouds when disturbed. Leaves mottled and covered with honeydew and sooty moulds where whitefly have sucked sap (see honeydew). **Control:** spray with Diazinon or Malathion. **11** Grey mould, caused by the fungus *Botrytis cinerea* (see Botrytis). May attack established plants. **Symptoms:** flowers rot and are covered with grey mould. **Control:** spray early with Benomyl. **12** Leaf-fall premature (see 18 below Premature leaf-fall). **13** Leafhopper, Rhododendron leafhopper *Graphocephala coccinea*. May attack rhododendrons at all stages. **Symptoms:** this large plant bug, torpedo-shaped and brightly coloured with blue, red and yellow markings, cuts slits into the flower buds to lay its eggs. This damage is thought to increase the spread of bud blast disease. **Control:** where bud blast is a problem, spray with Dimethoate, gamma-HCH, Malathion or Nicotine at fortnightly intervals from August to October to limit its spread by leafhoppers. **14** Leaf miner (see 1 above Azalea leaf miner). **15** Leaf scorch, caused by the fungus *Septoria azaleae*. May attack established plants. Attacks azaleas. **Symptoms:** brown patches on the older leaves; leaf margins thicken, and leaves fall when half the surface is affected; plants are weakened and flowering is poor. **Control:** spray with a copper fungicide. Varieties vary in susceptibility. **16** Leaf scorch and dead buds. May be physiological (see Physiological disorders), or may be caused by various species of fungi. May affect established plants. Rhododendrons chiefly affected. **Symptoms:** brown marks on the leaves; buds brown and may be dead. This condition is often associated with adverse weather conditions, such as frost, or very wet weather, or with damp ground. The dead buds rot and then may be attacked by botrytis (see Botrytis). There may also be direct fungal attacks on buds and leaves. **Control:** cut off badly affected and dead buds; open up growing conditions, if this is possible, to allow more air movement, but avoid too much exposure to wind; if nearby hedges or fences tend to create a wind pocket, consider what might be done to correct this, especially where early-flowering species and varieties are concerned. **17** Leaf spot (see Leaf spot). May be physiological (see Physiological disorders) or may be caused by various species of fungi. May affect established plants. Rhododendrons mainly affected. **Symptoms:** leaf spotting, with fungal disease present, which is generally secondary to damage caused by frost or sun scorching; leaf spotting with no disease present, due to weather damage. **Control:** improve growing conditions and, in the case of large-leaved species, provide shelter from strong sun, especially first thing in the morning. **18** Premature leaf-fall, dieback and brown buds. May be physiological (see Physiological disorders), or may be caused by various species of fungi. May affect established plants. **Symptoms:** leaves fall, shoots die back and buds turn brown. May be due to weather conditions such as frost, or in certain cases to unfavourable conditions during the transport of plants, such as damage, over-heating, chilling or dampness. *Azalea indica* plants which are packed and transported to growers for commercial glasshouse forcing at Christmas, New Year and Easter time, are often affected in this way. After the initial damage, fungal diseases may attack. **Control:** check, and if necessary, improve growing conditions; check glasshouse temperatures used for forcing; if possible, check conditions during transit. **19** Rhododendron bud blast, caused by the fungus *Pynostysanus azaleae*. May attack young and established plants. **Symptoms:** bud scales begin to turn black in autumn and later turn grey, with a covering of small black fruiting bodies of the fungus, and finally look quite black; the buds remain firmly attached, unlike buds damaged by frost which can be easily broken off. This disease is associated with rhododendron leafhoppers (see 13 above Leafhopper). It is thought that the leafhoppers damage buds when egg laying, opening up wounds to disease attack. **Control:** spray with insecticide to control leafhoppers; pick off and destroy all diseased buds if possible; copper sprays are useful. **20** Rhododendron bug *Stephanitis rhododendri*. May also be called Rust (see 23 below Rust). May attack young and established plants. **Symptoms:** black bugs feeding on the leaves causing rusty marks or mottled wrinkled effect; eggs laid in late summer may be seen on the undersides of the leaves. **Control:** spray with gamma-HCH, Malathion or Nicotine; avoid spraying in hot sunshine; on a small scale, crush the bugs. **21** Rhododendron whitefly *Dialeurodes chittendeni*. May attack young and established plants. **Symptoms:** mealy white adults feeding on undersides of young leaves in early to mid-summer; honeydew and sooty moulds on the leaves (see Honeydew). **Control:** spray with Diazinon, Malathion or Nicotine. **22** Root rot, caused by the fungus *Armillaria mellea* (see Armillaria). May attack young and established plants. **Symptoms:** wilting and death; examination of stems will usually reveal white fungal strands or black fungal 'boot laces' characteristic of this disease. **Control:** remove and burn infected plants; treatment of the soil with formalin can be reasonably effective but if possible plant on fresh ground. **23** Rust, caused by the fungus *Chrysomyxa rhododendri* (see Rust). May attack established plants. **Symptoms:** powdery yellow patches on the undersides of the leaves. This disease is unsightly rather than dangerous, so it is seldom necessary to take remedial action. **24** Scale, Soft scale *Coccus hesperidum* (see Scale insects).

May attack at all stages. Attacks evergreen plants. **Symptoms:** pale green to black, flattened scales are found in rows near to large veins on the undersides of leaves; scales produce honeydew leading to outbreaks of sooty moulds (see Honeydew). **Control:** to prevent the spread of scale insects, scrape off the plant, but if this is done, the affected area must be painted or sponged *at once* with a dilute solution of insecticide such as Diazinon, Malathion or Nicotine; on evergreen plants, incorporate small amounts of nicotine into leaf-shining oils which is more effective than wet sprays. **25** Thrips (see 9 above Glasshouse thrips). **26** Vine weevil *Otiorhynchus sulcatus*. May attack at all stages. Attacks Rhododendrons. **Symptoms:** adult weevils, black and 8 mm ($\frac{1}{3}$ in) long, feeding at night on leaf edges, producing a notched effect. **Control:** lightly incorporate gamma-HCH dust into the soil surrounding the plants; for plants in pots, apply banding grease in rings around the pots to provide an effective barrier to the adult which cannot fly when it tries to lay eggs on the plants. **27** Weevil (see 7 above Clay-coloured weevil, 26 above Vine weevil). **28** Whitefly (see 2 above Azalea whitefly, 21 above Rhododendron whitefly). **29** Wilt, Azalea wilt, caused by the fungus *Cyclindrocarpon radicicola*, and Rhododendron wilt, caused by the fungus *Phytophthora cambivora* (see Wilts). May attack established plants. **Symptoms:** wilting, drooping and withering of the leaves, due to the fungus clogging the conducting tissue in the stem; staining and distortion of the stem at ground level where the disease attacks. **Control:** many of the newer varieties are now grafted on to resistant stocks (see Resistant stocks), so use varieties with resistance to the disease; with greenhouse plants, avoid too deep planting or potting, and over-watering.

Rhubarb Varieties derived from *Rheum rhaponticum* and other *Rheum* species. Hardy herbaceous perennial, with woody, rhizomatous roots. Height: medium. Leaves: large with thick, fleshy stalks. Stalk colours: green or red. Use: food crop, grown for the fleshy leaf stalks. Grown in the open or forced for early rhubarb.

Pests and Diseases. 1 Crown rot *Erwinia rhapontica* (see Bacterial diseases). May attack established plants. **Symptoms:** purple leaves and poor growth; rotting of terminal buds and collapse of plants. This disease is often associated with various fungal diseases and with attacks by stem eelworm (see 4 below Stem eelworm). **Control:** there is no control other than lifting and burning diseased plants, and planting healthy new plants on fresh ground. **2** Grey mould, caused by the fungus *Botrytis cinerea*. May attack established plants. **Symptoms:** soft rotting and mould on stems. This disease does not affect rhubarb in the open except in very wet seasons, but may be troublesome on forced rhubarb. **Control:** improve ventilation and general conditions to reduce damage after an attack; various fungicides may be used regularly on forced rhubarb. **3** Leaf spot (see Leaf spot) caused by various species of fungi. May attack established plants. **Symptoms:** numerous brown spots on the leaves. **Control:** plants are not noticeably affected and no control is needed. **4** Stem eelworm *Ditylenchus dipsaci* (see Eelworms). May attack established plants. Often associated with Crown rot (see 1 above Crown rot). **Symptoms:** distortion of leaves and poor growth. This eelworm attacks many different kinds of plants. **Control:** lift and destroy affected plants; plant healthy new plants on fresh ground. **5** Virus, various virus diseases (see Virus diseases). May attack established plants. **Symptoms:** mottling of leaves and poor growth. **Control:** lift and destroy affected plants. Considerable efforts have been made by research workers to raise stocks of virus-free plants, and these are available in most areas.

Root rot A general term used to describe rotting of the roots of herbaceous plants, trees and shrubs, caused by fungal or bacterial diseases. In seedlings and young plants, root rot is usually known as damping off (see Damping off). Root rot is often associated with foot rot or stem rot (see Foot rot) and tends to be particularly bad at propagation time. Many root rots are secondary, following on, for instance, when plants have been growing in very unsuitable soils, or have suffered mechanical damage.

Control: discard severely affected plants; improve general growing conditions and use chemicals where possible to save less severely affected plants; prevent root rots by preparing ground properly for plants; maintain correct growing conditions throughout the life of plants.

Rootstocks (see Resistant rootstocks, Stock)

Rose Rose varieties have a parentage much more complicated than that of most cultivated plants. Modern rose varieties are very complex hybrids (see Hybrid), and for convenience, are put into various groups, such as climbing roses, floribunda roses, hybrid tea roses, miniature roses and others. A few *Rosa* species and varieties are grown, usually being called species roses or shrub roses. Hardy

Rose **Blackspot** caused by the fungus *Diplocarpon rosae*, early stage, blackish spots on upper surfaces of leaves (see page 193)

Rose **Powdery mildew** caused by the fungus *Sphaerotheca pannosa*, white powdery mould on leaves, thorns and buds (see page 195)

Strawberry **Mildew** caused by the fungus *Sphaerotheca macularis*, fruit covered in greyish powdery growth (see page 208)

Tomato **Blossom end rot** a physiological disorder, sunken, circular dark patches on the bottom end of the fruit (see page 215)

Tomato **Iron deficiency** a
physiological disorder,
yellowish blanching of leaves
(see page 218)

Tomato **Leaf mould** caused
by the fungus *Cladosporium
fulvum*, brown velvety spots
on underside of leaf (see page
219)

Tulip **Fire** caused by the
fungus *Botrytis tulipae*, spots
on flowers and grey mould on
leaves (see page 223)

Willow **Brassy beetle**
Phyllodecta species, small
metallic-coloured beetle (see
page 233)

shrubs, chiefly deciduous. Height: dwarf to large, climbing. Flowering time: summer in the open, other periods in the greenhouse. Flower colours: cream, lilac, lilac-blue, orange, pink, purple, red, white, yellow or parti-coloured. Flowers often very fragrant. Foliage: decorative, light green, dark green or coppery, and occasionally scented. Stems: spiny. Spines: may be ornamental, the hips varying in size and shape, and orange, red or yellow in colour. Use: beds, borders, trained to arches, fences, pergolas and walls, rock gardens, rose gardens, containers outside, greenhouse border and pot plants, commercial glasshouse cut flowers.

Pests and Diseases. 1 Aphis, Rose aphid *Microsiphum rosae*. May attack at all stages. **Symptoms:** pink-green or yellow-green aphids clustering on young leaves and buds. **Control:** spray with Oxydemeton-methyl, Dimethoate or Malathion. **2** Bee, Common leaf cutter bee *Megachile centuncularis*. May attack young and established plants. **Symptoms:** neat little incisions taken out of the edges of leaves. This is not a serious pest. So no control is necessary. **3** Black mildew (see 11 below Downy mildew). **4** Blackspot, caused by the fungus *Diplocarpon rosae*. May attack at all stages. **Symptoms:** blackish or purplish spots, irregular in size, are first visible on the upper surfaces of the leaves; as the spots grow larger, they develop yellow edges; finally the leaves wither and drop. Severe attacks weaken the plants. This may be a serious disease in clean air areas, less so in industrial towns. **Control:** collect and burn diseased leaves; spray regularly every fortnight with Captan, Dichlofluanid, Maneb (in combined form) or Zineb; spray the ground in early spring with a colloidal copper spray or a household disinfectant; mulch in spring with peat, spent hops or even grass cuttings, provided these are not from a lawn that has been treated with hormone weedkillers. Nurserymen's catalogues usually list varieties resistant to black spot. **5** Canker (see 21 below Stem canker). **6** Capsid bugs *Lygus* species (see Capsid bugs). May attack at all stages. **Symptoms:** small punctures on young leaves, followed by distortion which may be acute; buds may turn brown. **Control:** at first sign of attack, spray with Diazinon or Malathion. **7** Caterpillars, caterpillars of lackey moth, tortrix moth and others (see Caterpillars). May attack at all stages. **Symptoms:** large areas eaten out of the leaves; caterpillars visible on the plants; in severe attacks plants may be defoliated. **Control:** spray with Carbaryl, Derris (Rotenone), Malathion or Trichlorphon; if necessary, vary the sprays at intervals to ensure effective control. **8** Chlorosis (see Physiological disorders). May occur at all stages. **Symptoms:** leaves become yellow and mottled. This is due initially to excess lime in the soil, which has the effect of locking up available iron. **Control:** apply iron sequestrine material (see Sequestrols) to the soil or, in severe cases, flowers of sulphur; mulch with sphagnum peat which may help. **9** Cuckoo-spit bug (see Cuckoo-spit bugs). **10** Dieback. May be caused by various species of fungi or may be physiological (see Physiological disorders). May attack young or established plants. **Symptoms:** plants gradually die. **Control:** seek specialist advice to establish the cause of dieback; remove and burn affected plants; plant new ones on fresh ground, or at least in fresh soil. **11** Downy mildew, Black mildew, caused by the fungus *Peronspora sparsa*. May attack young and established plants. **Symptoms:** leaves turn brownish-purple on the upper surface and greyish-white on the reverse; leaves shrivel and fall; young shoots and flower stalks may also be attacked. This is a serious disease, more common in glasshouses than in the open. **Control:** collect and burn all fallen leaves; spray at regular intervals with copper/petroleum oil emulsion or Zineb; avoid high humidity and condensation at night by giving adequate ventilation and warmth. **12** Grey mould, caused by the fungus *Botrytis cinerea* (see Botrytis). May attack established plants. **Symptoms:** buds fail to open, and look mummified, or buds open, and flowers split; branches die back. This is a general disease which may attack many plants, and in wet summers, roses may be badly affected, especially if they are growing in bad situations with no air movement. **Control:** spray with Benomyl, but it is also important to open up the bushes well when pruning and to reduce nitrogen fertilizers when feeding; it may be necessary to move plants to a better position. **13** Leafhopper, Rose leafhopper *Edwardsiana rosae*. May attack at all stages. **Symptoms:** yellow larvae hatch from eggs laid beneath the surface of the leaves and on the undersides of leaves, causing mottling; moult skins may be seen on the leaves. **Control:** spray regularly with Derris, Malathion or other insecticides. **14** Leaf spot, caused by the fungus *Sphaceloma rosarum* and also by *Phyllosticta* and *Septoria* species of fungi (see Leaf spot). May attack established plants. **Symptoms:** small black spots on the leaves; severely attacked leaves turn red and may fall; red-brown scales appear on the stems. **Control:** as for Black spot (see 4 above Blackspot). **15**

Mildew (see 11 above Downy mildew, 17 below Powdery mildew). **16** Mosaic (see 24 below Virus). **17** Powdery mildew, caused by the fungus *Sphaerotheca pannosa* (see Mildew). May attack young and established plants. **Symptoms:** leaves and tips of shoots covered with white powdery mould, which will spread over the thorns and buds. This is a serious disease. **Control:** spray regularly with Benomyl, Dinocap, lime sulphur or commercially with Quinomethionate; free movement of air round plants is very important, and this should be considered in planning planting distances, and in pruning and general care. **18** Rust, Rose rust, caused by the fungus *Phragmidium mucronatum* (see Rust). May attack established plants. **Symptoms:** orange pustules on leaves during summer, with darker brown pustules later in the season. This is a serious disease of roses in the open. It seldom occurs in greenhouses. **Control:** remove and burn infected leaves; spray with Mancozeb/Zineb commercially, Maneb or Thiram; burn winter prunings. **19** Sawfly. (a) Banded rose sawfly *Allantus cinctus*. May attack established plants. **Symptoms:** green larvae tunnel into branches. **Control:** cut out branches below the level of attack; spray with Derris or gamma-HCH. (b) Leaf rolling rose sawfly *Blennocampa pusilla*. May attack established plants. **Symptoms:** leaves rolled up. The habits of rolling up leaves makes this pest difficult to control. **Control:** pull off and burn attacked leaves, and spray with gamma-HCH and Malathion. (c) Rose slug sawfly *Endelomyia aethiops*. May attack established plants. **Symptoms:** larvae, which are yellow and slug-like, feed on the leaves and skeletonize them. **Control:** spray with Derris or gamma-HCH. **20** Scale, Rose Scale *Aulacaspis rosae*. May attack established plants. **Symptoms:** stems are spotted with white circular or oblong scales, with red-brown larval skins in the centres. **Control:** spray with Diazinon or Malathion at 21-day intervals. **21** Stem canker, caused by the fungus *Leptosphaeria coniothyrium*. May attack young and established plants. **Symptoms:** red-brown or purplish areas on young green wood of stem which later form cankers with rough irregular margins; darker brown lesions and cankers may also occur at areas of grafting. **Control:** cut away and burn diseased tissue; use proprietary paint to protect wounds; if grafting, take care to leave no unprotected wounds as infection can enter through wounds; spraying with fungicides is of little use. **22** Thrips, Rose Thrips *Thrips fuscipennis*. May attack established plants.

Symptoms: leaves, shoots and buds become distorted, and show small black markings. **Control:** spray with Derris, gamma-HCH or other general insecticide. **23** Verticillium wilt, caused by the fungus *Verticillium albo-atrum* (see Verticillium). May attack established plants. **Symptoms:** wilting of shoots. **Control:** remove and burn affected bushes. Not a common disease of roses. **24** Virus, Rose mosaic virus and other virus diseases (see Virus diseases). **Symptoms:** mottled and distorted leaves. **Control:** remove badly affected bushes. **25** Weedkiller damage (see Physiological disorders). **Symptoms:** distorted and puckered leaves. This is due to drifting of hormone weedkillers, or to the use as a mulch of lawn cuttings when the lawn has been treated with hormone weedkillers. Other weedkillers might be responsible. Roses may recover in time, but they often die. **Control:** check on the use of all weedkillers in the area, and use weedkillers with care. Note that, as with all plants, various factors such as climate, condition of soil, quality of plant, planting methods, use of fertilizers and manures and general cultivation, affect the normal growth of roses, and if unfavourable, may make them more liable to pest and disease attack.

Rose of Sharon   (see St John's wort)

Rotation   A system of periodically changing the crop in the same ground. Land which is used year after year for the same plants tends to build up pests and diseases, and micro-organisms, specific to those plants or to plants of similar botanical grouping. For example, where brassicas are grown continually on the same land, club root disease becomes a major problem, or again, if tomatoes are grown for more than one or two years in unsterilized soil, they will rapidly build up root troubles. In addition, land becomes 'sick' of the same crop, due to exudations of plant acids, and other factors. There are also nutritional considerations in crop rotation as different plants prefer different levels or a different balance of food. In the vegetable garden, a 3-year rotation of crops is advisable and further information on vegetable culture is obtainable from any book dealing with the subject. Other types of plants may also need rotation. Strawberries, for instance, should not be grown in the same soil for more than 3–4 years. A long rotation is advised where there have been pests or diseases which tend to persist in the soil.

Rubber plant   (see Ficus)

Russian vine   (see Cissus)

Rust   A term often used to describe any sort of brown, orange or red discolouration on plants, but it should be confined to the rust diseases, to avoid confusion. These are a large group of fungal diseases (see Fungal diseases) which live as parasites on various plants, existing on dead organic matter in the resting spore stage only. Some rust diseases are confined to one species of plant only, others have the habit of spending part of their life cycle on one species of plant, and part on another. **Symptoms:** small, raised, orange, red or yellow spots on stems or leaves. **Control:** various chemicals may be used to control rusts.

S

St John's wort   *Hypericum calycinum* Rose of Sharon, usually called St John's wort, and other *Hypericum* species. Hardy deciduous and evergreen shrubs and sub-shrubs and herbaceous perennials. Height: herbaceous perennials, small; shrubs, small to medium. Flowering time: summer. Flower colour: yellow. Use: borders, dry shaded banks, rock gardens.
Pests and Diseases. 1 Rust caused by the fungus *Melampsora hypericorum* (see Rust). May attack young and established plants. **Symptoms:** orange pustules on the undersides of the leaves, which restrict growth considerably, especially on young plants struggling to get established. **Control:** spray with Maneb, Thiram or Zineb or commercially with Mancozeb.
Salts   (see Soluble salts)
Salvia   A genus of over 500 various herbs, sub-shrubs and shrubs. The most common species are *Salvia azurea, Salvia greggii, Salvia haematodes, Salvia leucantha, Salvia nutans, Salvia officinalis* sage, *Salvia patens, Salvia pratensis, Salvia sclarea* clary, *Salvia splendens* the summer bedding type, *Salvia virgata.* Hardy, half-hardy, herbaceous perennials, annuals and biennials. Height: small to medium. Flowering time: varies with sort and method of growing throughout the year. Flower colours: cream, blue, lilac,

pinkish, purple, red, white or parti-coloured. In some varieties the bracts are coloured and long-lasting. Foliage: variegated. Use: annual borders, herbaceous borders, summer bedding, outdoor containers, window boxes, greenhouse pot plants.

Pests and Diseases. 1 Glasshouse leafhopper *Hauptidia maroccana*. May attack at all stages. **Symptoms:** white or yellow spots on foliage extending to form bleached areas. The adults are visible, slender and angular, pale yellow, with dark V-bands on the back. **Control:** fumigate with gamma-HCH at 14 to 28-day intervals. **2** Glasshouse whitefly *Trialeurodes vaporariorum* (see Whitefly).

Sansevieria *Sansevieria trifasciata* and its *laurentii* variety, and *Sansevieria hahnii*. Greenhouse herbaceous perennials with thickened rhizome. Height: medium to large. Leaves: decorative, long, narrow, flat, thick and with sharp tips. Leaf colour: variegated. Use: foliage plant, greenhouse border and pot plant, indoor pot plant.

Pests and Diseases. 1 Leaf spot, caused by the fungi *Fusarium monilforme* and *Gloeosporium sansevieriae*. May attack established plants. **Symptoms:** brown markings on the leaves. **Control:** spray with Bordeaux mixture; keep plants in warm conditions. **2** Rolling of the leaves (see Physiological diseases). **Symptoms:** rolling inwards of the upright, sword-like leaves. This is usually due to cold conditions, but disease infection may make it worse (see 1 above Leaf spot). **Control:** if disease is present, raise temperature.

Savoy Varieties derived from the cabbage *Brassica oleracea* (see Cabbage). Hardy biennial, with stem woody at the base. Height: small to medium. Foliage: leaves waved or blistered, forming large, compact, pointed or rounded heads. Use: food crop, grown for the leafy heads. For **Pests and Diseases**, see Brassicas.

Scale insects Scale insects are related to mealybugs (see Mealybugs), but unlike them, are inactive for most of their lives. They develop a scaly covering, which gives them their name. Various species are serious pests of a wide range of plants in the greenhouse and in the open, and at all stages of growth. The young scale insects swarm over the plants, making them look as if they were covered with powder. Scale insects suck the plant sap for food and excrete a sticky substance called honeydew. Black moulds, called sooty moulds, tend to grow on the honeydew, and this dark sticky substance makes the plants unsightly and interferes with their normal growth. As young scale insects develop they lose their legs and become covered with scales which may be soft and flat or rounded and hard according to species. These scales remain closely attached to the plants, disfiguring and weakening them. It is easier to destroy young scale insects than adults.

Control: use one of the various proprietary sprays available at the first sign of attack and repeat as necessary; take routine pest control measures in the greenhouse; in winter, spray fruit trees and bushes outside to help control scale insects and use a suitable winter spray on ornamental trees and shrubs outside, if attacked.

Seed dressings   Seed dressings are used to disinfect seed and protect it from soil organisms which may attack it, especially in cold, damp conditions. Seed is shaken up with various chemicals, for example, organic mercury dust. Seed dressings are particularly useful for seed sown in early spring or late autumn, such as seed of brassicas, beans and peas. Seed may also be soaked in fungicide, for example, Thiram. Seedsmen may supply treated seed. The treatment has to be done correctly or there will be injury to the seed which may check or even prevent germination. Dressed seed should never be left on the soil surface in case they are eaten by any wild birds.

Seed treatment   (see Seed dressings)

Sequestrols   Substances used in the treatment of mineral deficiencies in plants (see Mineral deficiencies) especially deficiencies of iron and magnesium. They consist of trace elements, which are elements essential to the normal growth of plants but, unlike the major elements, are required only in extremely minute quantities. Sequestrols are prepared in a form that enables plants to take them up through their roots from soils, usually alkaline soils, that otherwise would hold the elements and prevent plants from absorbing them. Sequestrols are also known as sequestrenes or as sequestrated or chelated compounds.

Silver firs   *Abies fargesii*, *Abies grandis* Giant Fir, *Abies homolepis* Nikko fir, *Abies nordmanniana* Caucasian fir and other *Abies* species and varieties. Hardy, evergreen, coniferous trees and shrubs. Height: shrubs, prostrate to large; trees, small to large. Variable habit of growth. Foliage: may be gold or white, variegated. Cones: erect and usually cylindrical, coloured violet-blue when young in some species. Use: planted singly or in groups or avenues in large areas, conifer borders, windbreaks, rock gardens.

Pests and Diseases. 1 Aphis (see 2 below Green spruce aphid, 5 below Spruce gall adelges). **2** Green spruce aphid *Elatobium abietinum*. May attack at all stages. **Symptoms:** excessive needle drop preceded by yellowing of needles. **Control:** spray with Malathion in March or April if practical. **3** Rust, caused by the fungus *Melampsorella caryophyllacearum* (see Rust). May attack at all stages. **Symptoms:** orange pustules on the needles; general poor growth. **Control:** size of tree may make fungicidal spraying impractical, but use Thiram and Zineb which are effective on small trees. **4** Scorched needles, caused by the fungus *Rehmiellopsis bohemica*. May attack at all stages. **Symptoms:** red needles in spring, turning brown/black and shrivelling later; in serious cases, branches die. **Control:** on young trees try fungicides such as Benomyl or Captan, otherwise none known. **5** Spruce gall adelges, Woolly aphid *Adelges abietis*. Attacks are most severe on trees over four years old. **Symptoms:** woolly patches on the needles in spring followed by development of pineapple-shaped galls. **Control:** spray with gamma-HCH plus wetting agent in March and remove any galls developing in summer. **6** Woolly aphid (see 5 above Spruce gall adelges).

Slaters   (see Woodlice)

Slugs and snails   Slugs and snails are familiar, widespread and destructive pests. Slugs are soft and legless and, according to species, vary in size, and in colour from black to grey. Snails are similar but smaller and have the protection of spiral shells, varying in size, colour and patterning according to species. Snails are not as abundant or as destructive as slugs. Both slugs and snails excrete a slimy substance which leaves trails on soil or plants. This slime helps them to move freely and keeps them free of dust, grit and any material that may fall on them. Clusters of white, translucent eggs are laid in the soil. Young slugs and snails at first feed on old organic matter, but soon begin to attack many different kinds of plants, especially those with soft leaves and stems. They usually feed by night, different species attacking plants below ground level, at ground level or higher up on the plants. By day, they hide in suitable dark moist places. They feed throughout the year, unless the weather is very severe in winter or very hot in summer, when they hide until conditions are favourable for them to come out and start feeding again. Some common species are the black slug *Arion ater*, the field slug *Agriolimax reticulatus*, the garden slug *Arion hortensis*, the keeled slug *Milax budapestensis* and the white-soled slug *Arion circumscriptus*, and the banded snails *Cepaea hortensis* and *Cepaea nemoralis*, the garden snail *Helix aspersa* and the strawberry snail *Hygromia striolata*.

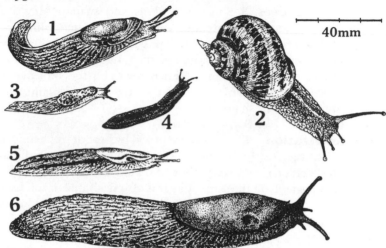

40mm

*1 Milax budapestensis 2 Helix aspersa 3 Agriolimax reticulatus
4 Arion hortensis 5 Arion circumscriptus 6 Arion ater*

Control: avoid excessive use of organic matter, especially where slugs and snails are known to be abundant; avoid leaving weeds or other organic matter lying on the ground or in haphazard heaps; use poison baits (see Poison baits), with traps of orange or grapefruit peel and pieces of slate; use Agricultural salt or put old ashes or other gritty material around plant roots.

Snails (see Slugs and snails)

Snowdrop *Galanthus nivalis* Snowdrop and varieties, and other *Galanthus* species. Hardy herbaceous perennials, with bulbs. Height: dwarf to small. Flowering time: autumn to spring according to sort. Flower colour: white/green. Use: mainly naturalizing in grass or open woodland, also rock gardens, commercial and garden cut flowers, short-term indoor pot plants, cold greenhouse pot plants.

Pests and Diseases. 1 Grey mould, caused by the fungus *Botrytis galanthina* (see Botrytis). May attack established plants. **Symptoms:** young shoots covered with downy growth which eventually turns brown. **Control:** spray with Benomyl; as attacks may follow poor growth caused by cold winds or by badly drained soil, improve growing conditions. **2** Mice, Field mouse *Apodemus sylvaticus*, House mouse *Mus musculus*. May attack established plants. **Symptoms:** bulbs eaten. **Control:** if necessary and if possible, use traps. **3** Slugs (see Slugs and snails). May attack established plants. **Symptoms:** young shoots eaten; slimy trails visible. **Control:** if necessary and if possible, put down poison baits (see Poison baits). **4** Smut, caused by the fungus *Urocystis galanthi*. May attack established plants. **Symptoms:** blisters on leaves and flower sheaths. **Control:** lift and destroy affected plants; plant healthy bulbs on fresh ground.

Soil sterilant   (see Soil sterilization)

Soil sterilization   The treatment of soil by heat or by chemicals to kill pests, weed seeds and the fungal spores which spread diseases. Soil sterilization is necessary when crops such as tomatoes are being grown repeatedly in the same soil. It is also desirable in soil used in seed trays or for potting plants. The standard commercial method is by the application of steam using special equipment, but this is economical only where large areas are to be treated each year. Small quantities of soil may be sterilized by heat, placing the moist soil in a domestic oven at 100°C (212°F) for at least half an hour. Electric sterilizers are also available. These operate by dry heat or by the production of steam which passes through the soil. Of the chemical treatments, few effective ones are available to the amateur. Basamid (Dazomet) is available in 5 kg packs. The chemical releases a toxic gas into the soil and can be used in late autumn and winter outdoors or at any time under glass or in potting composts. Covering the soil with polythene to contain fumes is helpful and improves the effectiveness of the sterilant. Careful soil preparation is important and ideally, a rotary cultivator should be hired to incorporate the material through the soil, although thorough forkings can be reasonably effective. Glasshouse soils can be sterilized by the use of methyl bromide, a very dangerous gas, which can only be applied by qualified contractors. Other chemicals such as Formalin have limited value in controlling soil disorders but can be used to sterilize implements, split canes and the like.

Soil testing   For successful growth, all soils and growing media (see Composts) for plants, must contain adequate nutrients and must be of a suitable pH (see pH). The amount of nutrients necessary will vary according to type of plant. Soils are accurately tested for their pH and nutrient content in laboratories, but the gardener may gain a fairly accurate idea of soil pH and of nitrogen, potassium and phosphorus content of the soil by using a soil testing kit, which depends on colour comparisons. Various types of kit are available and, if directions are followed, results can be remarkably accurate, and are certainly adequate for most amateur gardening activities. It should be remembered that soils and composts containing too high a level of plant nutrients can be toxic to plants, causing damage to the roots and even to the whole plant.

Soluble salts   Certain chemical substances essential to the normal growth of plants are present in solution in the soil, and are taken up by plant roots. The chemical substances present may be supplemented and their intake influenced by adding fertilizers, lime and manures to the soil. When the soluble salts present in the soil, or fertilizers which have been added, are in excess of what plants require, growth may be greatly restricted or actually damaged by the excess concentration of soluble plant foods (see Osmosis).

Species   Term for a group of individual plants having the same characteristics, which tend to remain constant, though they may be modified. Species are contained in a genus (see Genus).

Spinach   Derived from *Spinacia oleracea*. Hardy annual. Height: medium. Leaves: large and fleshy. Use: food crop, grown for its leaves.

Pests and Diseases. 1 Aphis, Black bean aphid and Blackfly *Aphis fabae* (see Aphis). May attack young and established plants. **Symptoms:** deterioration and crippled growth if attack is severe. **Control:** spray with Derris. **2** Downy mildew caused by the fungus *Peronospora effusa* (see Mildew). May attack established plants. **Symptoms:** yellowish spots on leaves with mauvish mould on the undersides of the leaves; badly affected growth. This disease may be particularly bad in wet weather, on damp ground and where growth is too soft, owing to excess use of nitrogen manures and fertilizers. **Control:** spray with Zineb, but in future improve growing conditions. **3** Leaf miner, similar to Mangold fly (see **5** below Mangold fly). **4** Leaf spot, caused by the fungus *Heterosporium variabile*. May attack established plants. **Symptoms:** round brown sunken spots on the leaves may increase until the leaves are dried up and destroyed. **Control:** spray with a copper fungicide or Zineb. **5** Mangold fly *Pegomya hyoscyami*. May attack at all stages. **Symptoms:** blistered and shrivelled leaves;

young plants may be killed. Eggs are laid in spring and larvae tunnel between the two skins of the leaves, causing blisters. There may be 3 generations in the year. **Control:** spray with Trichlorphon. Commercial growers may use Mevinphos or other chemicals. **6** Slugs (see Slugs and snails). **7** Virus, Spinach blight (see Virus diseases). **Symptoms:** yellowing of the leaves and dwarfing of the plants. This disease is spread by aphids. **Control:** guard against aphids.

Spindleberry   (see Euonymus)

Springtails   This name is given to species of *Collembola*, minute, colourless, dark or white wingless insects, which move actively and

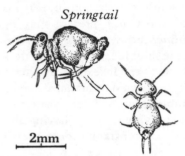

Springtail

2mm

spring by means of their tails. They feed on dead and dying organic matter, but may attack plants, bulbs, roots or tubers already damaged by other pests or by disease. In large numbers they can be very destructive in glasshouses. Seedlings and young plants show pin holes in the leaves and scraping of the surface. To find out if springtails are present, float some soil from around the plants in a bucket of water. If present, springtails will rise to the surface in large numbers.

Control: apply a gamma-HCH dust to the soil, or drench with gamma-HCH or Malathion.

Spruce   *Picea abies* common spruce or Christmas tree and varieties, and other *Picea* species and varieties. Hardy, evergreen, coniferous, trees and shrubs. Height: dwarf and prostrate shrubs to large trees. Cones: drooping, green, lilac, purple or red when young, brown when mature. Foliage: attractive, varying in colour and form, according to species and variety. Growth: varied and interesting. Use: planted singly or in groups or rows in large gardens, parks or open spaces, or in mixed conifer borders; dwarf and prostrate sorts in beds, borders and rock gardens, and as ground cover.

Pests and Diseases. 1 Bud blight, caused by the fungus *Cucurbitaria picea*. May attack young and established plants. **Symptoms:** blackened tips of shoots and buds; black spores on affected buds. **Control:** spray young and nursery trees with a copper fungicide. Little can be done for larger trees. **2** Butt rot, caused by the fungus *Fomes annosus*. May attack young plants.

Symptoms: sudden death of trees 5–15 years old; exudation of resin at the base of trunks, usually followed by small flat fungal bodies; rotted centres of trees. **Control:** seek specialist advice from an arboriculturist, and ask about resistant species of firs and larches; check that growing conditions are adequate and that soil is not too shallow. **3** Cone rust, caused by the fungus *Thekopsora areolata*. May attack established plants. **Symptoms:** brown spores which turn yellow, covering cones and causing cones to remain open all the time. **Control:** no control necessary or practical. **4** Grey mould, caused by the fungus *Botrytis cinerea*. May attack young plants. **Symptoms:** poor growth; grey mould visible on shoots. In wet seasons, severe damage may be caused to young trees, especially in the nursery. **Control:** spray with Benomyl; improve growing conditions in the nursery. **5** Leaf scorch, caused by the fungus *Lophodermium macrosporum*. May attack young or established plants. **Symptoms:** a red hue at almost any time of the year on trees; falling needles. It is difficult to do anything for larger trees. **Control:** spray young trees in the nursery with a copper fungicide or Maneb. **6** Needle rust, caused by the fungus *Chrysomyxa abietis*. May attack established plants. **Symptoms:** pale bands on the needles, which later turn yellow and drop. Only young needles are affected. **Control:** generally, no control necessary, or indeed practical on larger trees. **7** Pine beetle *Myelophilus piniperda*. **Symptoms:** adult beetles and larvae boring under tree bark, making tunnels; young tree shoots falling off; stunted growth and deformed crowns. Trees in poor condition and cut wood lying on the ground attract the beetles. **Control:** cut down and remove unhealthy trees and logs. **8** Pine weevil *Hylobius abietis* (see Weevils). May attack young and established plants. **Symptoms:** brown or black adult weevils in buildings near trees, as well as on trees; eggs are laid in fallen wood or logs, and larvae burrow in the wood; adults later attack trees by feeding on young shoots and can kill young trees. **Control:** clear ground of fallen branches, dead wood and logs; spray ground with gamma-HCH. **9** Rhododendron rust, caused by the fungus *Chrysomyxa rhododendri* (see Rust). May attack established trees. **Symptoms:** needles affected. Infection spreads from rhododendrons and control on spruce is difficult. **Control:** remove rhododendrons if practical. **10** Weather damage (see Physiological disorders). **Symptoms:** reddening or yellowing of needles. Reddening due to cold winds, yellowing to starvation.

Trees in plantations are less affected by cold than trees growing naturally. **Control:** apply a slow-acting nitrogen fertilizer to improve colour. **11** Witches' brooms (see Physiological disorders). Various fungi accompany this disorder. **Symptoms:** dense leafy clusters of shoots. **Control:** cut off the brooms and paint over cuts with tar oil wash.

Stachys   *Stachys lanata* Lamb's tongue, *Stachys macrantha*, *Stachys coccinea* and other *Stachys* species and varieties. Hardy herbaceous perennials. Height: small. Flowering time: summer. Flower colours: pink, purple, red or parti-coloured. Some grown for the decorative value of their large, woolly-white or grey leaves. Use: herbaceous borders, summer bedding, greenhouse pot plants. **Pests and Diseases. 1** Leaf spot, caused by various fungi (see Leaf spot).

Stem rot   A general term used to describe diseases which cause rotting of plant stems at or near soil level. It is usually applied to fungal diseases of herbaceous plants. It is also known as foot rot (see Foot rot). Stem rot and leaf spot are often associated (see Leaf spot).

Stock   This term has various meanings: (a) a collection or race of plants, kept so that supplies can be provided at any suitable time; (b) the crown of a plant, from which roots and shoots emerge; (c) the rooted part of a plant to which buds or grafts are applied. It is usually the stem which is budded or grafted, but it may sometimes be an underground stem part, such as a rhizome, which is used for grafting, and this is called the rootstock. The term rootstock may also be used simply to mean the stock (root portion) on which the scion is grafted (see Grafting).

Strain   This term is used for a group of plants, raised from seed, from a particular cultivar or variety (see Cultivar, Variety), or a group of closely related cultivars or varieties which have been carefully selected on account of certain desirable qualities, for example, hardiness, disease-resistance, earliness or size, and have been bred to retain these qualities.

Strawberry   Varieties derived from a number of different species of *Fragaria*. Hardy herbaceous perennials with runners. Height: small. Flowering time: spring. Flower colour: white. Fruit: creamy-pink, crimson, or red. Use: food crop, grown for the fruit; may be grown in pots in the greenhouse on a small scale. Where plants in the greenhouse may be attacked, this is noted. **Pests and Diseases. 1** Aphis, Shallot aphid *Myzus ascalonicus*,

Strawberry aphid *Chaetosiphon fragaefolii*. May attack at all stages.
Symptoms: leaves curled and distorted; pale yellow or greenish
aphids feeding on the plants. **Control:** spray with Demeton-S-
methyl commercially, Dimethoate or Malathion prior to flowering
and after picking is completed. **2** Birds, various species. May attack
and spoil fruit. **Control:** only satisfactory method is netting. **3**
Botrytis, Fruit rot, Grey mould, caused by the fungus *Botrytis
cinerea* (see Botrytis). **Symptoms:** soft rot of flowers and fruit,
which are covered with soft grey mould. This disease may cause a
severe loss of fruit, especially in wet seasons, in wet areas and where
ground tends to be damp. It may attack many different kinds of
plants, and its spores are everywhere. Spores overwinter on
strawberry plants, on plants near them and on the soil, and can start
infection in the spring. **Control:** to gain control of botrytis by
chemical means, take preventive measures *before* attack starts; use
Benomyl, Captan, Carbendazim, Dichlofluanid, Thiophanate-
methyl and Thiram, all of which are recommended chemicals;
while directions given with the product must be followed, basically
start all spraying at the flowering stage, repeating twice at 10 to 14-
day intervals; if strawing down is practised, spray immediately after
strawing; plastic mulching can help to control botrytis, as also can
wider spacing; Captan should not be used where fruit is intended for
canning; plants in the greenhouse may also be attacked. **Control:**
avoid over-damp conditions around the plants; spray with above
chemicals at 10-day intervals from early flowering to picking. **4**
Capsid, Common green capsid *Lygocoris pabulinus*. May attack
young and established plants. **Symptoms:** leaves punctured and
curled. **Control:** spray with Dimethoate as soon as leaf curling is
seen. **5** Carabid or seed beetle (see 24 below Strawberry seed beetle).
6 Chafer beetles (see Chafer beetles). **Symptoms:** plants
wilted by larvae feeding on the roots. **Control:** drench plants in
September with Carbaryl. **7** Cuckoo-spit bug *Philaenus spumarius*
(see Cuckoo-spit bug). May attack young and established plants.
Symptoms: crumpling and reddening of the leaves with
blackening of the veins. **Control:** spray with Malathion. **8**
Cutworms (see Cutworms). May attack young and established
plants. **Symptoms:** larvae eat into the roots and the stems at soil
level, causing considerable damage, and even death of plants.
Control: when damage is seen, apply Carbaryl or put down poison
bait (see Poison baits). **9** Eelworm (see 13 below Leaf and bud

eelworms). **10** Fruit rot (see 3 above Botrytis). **11** Glasshouse red spider mite *Tetranychus urticae* (see Tomato 18 Glasshouse red spider mite). May also attack plants in the greenhouse. **12** Grey mould (see 3 above Botrytis). **13** Leaf and bud eelworms, Red plant *Aphelenchoides fragariae*, *Aphelenchoides ritzemabosi* (see Eelworms). May attack young and established plants. **Symptoms:** small distorted leaflets with rough grey/brown areas on them; leaves are almost hairless and stalks are elongated and may turn red, or stalks may be very short; main growth may be killed. **Control:** always buy certified plants (see Certified stock); do not replant on eelworm-infested soil; use sterilized compost for plants grown in the greenhouse; it is possible to treat runners from infested plants by hot water treatment, soaking in water at 46°C (115°F) for 10 minutes (see Hot water treatment), but this is a specialized technique used by commercial growers. May also attack plants in the greenhouse. **14** Leatherjackets *Tipula paludosa* (see Leatherjackets). May attack young and established plants. **Symptoms:** wilting of plants due to larvae eating into roots or stems at ground level. This pest may cause considerable damage or complete death of plants. **Control:** apply Carbaryl mid-October to April; or put down poison bait (see Poison bait). **15** Mildew, caused by the fungus *Sphaerotheca macularis* (see Mildew). May attack established plants. **Symptoms:** dark areas on the upper surfaces of the leaves, with greyish powdery growth on the undersides; leaf margins curl upwards, exposing the undersides and reducing the efficient functioning of the leaves; flowers and fruit may be affected also. **Control:** spray with Benomyl, Dinocap, lime sulphur, sulphur or Thiophanate-methyl at 10 to 14-day intervals from just before flowering onwards; stop using lime sulphur when the fruit starts ripening. May also attack plants in the greenhouse. **16** Red core, caused by the fungus *Phytophthora fragariae*. May attack young and established plants. **Symptoms:** stunted plants with poor leaves, the older ones dying off; roots turn black and rot away; the outer skins of infected roots may be easily rubbed off, showing the central core which is red in colour. This is a serious disease. **Control:** while resistant varieties are available, no complete control is known, apart from long rotation (see Rotation); on a small scale, soil sterilization (see Soil sterilization) could be tried, using Basamid. **17** Red-legged weevil *Otiorhynchus clavipes* (see 29 below Weevils). **18** Seed beetle (see 24 below Strawberry seed beetle). **19**

Slugs and snails (see Slugs and snails). **20** Squirrel. May do very serious damage in certain areas by eating fruit. **Control:** only sure method is by netting plants. **21** Strawberry blossom weevil *Anthonomus rubi* (see Raspberry 3 Blossom weevil). **22** Strawberry mite *Tarsonemus pallidus*. May attack young and established plants. **Symptoms:** very small colourless mites feed on young, partly unfolded leaflets, making them small, wrinkled and brown; older leaves are silvery brown; plants are stunted and may die. **Control:** spray with Dicofol 3 times at 28-day intervals from the first signs of damage; remove and burn affected leaves. May also attack plants in the greenhouse. **23** Strawberry rhynchites *Caenorhinus germanicus* (see Weevils). May attack established plants. **Symptoms:** unopened buds shrivel and may fall. The adult weevils are about 2.5 mm ($\frac{1}{8}$ in) in length and of a metallic green/blue colour. They feed on the leaves, and when blossom starts to develop, on the flower stalks, which prevents buds from opening. May also attack raspberries. **Control:** spray with Carbaryl or HCH when attack is first seen. **24** Strawberry seed beetle *Harpalus rufipes*. May attack established plants. **Symptoms:** greyish active beetles about 12 mm ($\frac{1}{2}$ in) in length feeding on the seeds and flesh when the fruit is almost ripe, completely spoiling it. **Control:** difficult to control but promising results have been obtained by using a poison bait of 30 ml (1 fl. oz) Malathion to 1 kg (2 lbs) crushed oats. It is important to use rubber gloves when handling this bait, and to avoid dropping it on to fruit as it would cause taint. Methiocarb may also be used. **25** Strawberry tortrix moth *Acleris comariana*. May attack plants. **Symptoms:** caterpillars hatch in May, are green with darker green backs, about 6–7 mm ($\frac{1}{4}$ in) long when mature, and very active when disturbed; they feed on unopened leaves and later on buds and flowers; they weave leaves together with silken threads into tents as protection. **Control:** spray in April/May with Derris (Rotenone), Fenitrothion or Trichlorphon. **26** Tortrix moth (see 25 above Strawberry tortrix moth). **27** Vine weevil (see 29 below Weevils). **28** Virus (see Virus diseases). (a) Arabis mosaic. **Symptoms:** in late spring and autumn yellow spots or, in some varieties, bright red spots, or blotches appear on the leaves; distortion of the leaves and dwarfing may occur. **Control:** destroy affected plants; avoid replanting with runners from affected plants; buy new plants from a reliable source and plant on fresh ground. Commercial growers may control the eelworm *Xiphenema diversicaudatum*, which carries this virus, by DD injection in the autumn, or by soil sterilization with Basamid. (b) Crinkle, Green petal virus, Yellow edge. **Symptoms:** symptoms vary and include leaf discolouration and distortion, and stunting of the plants; with green petal virus, sepals (small leaves surrounding the flower) are enlarged and petals dwarfed, so that the flowers appear to be green. **Control:** destroy affected plants; control aphids as some of the viruses are carried by aphids; buy new plants from a reliable source. May also attack plants in the greenhouse. **29** Weevils, Red-legged weevil *Otiorhynchus clavipes*, Strawberry blossom weevil *Anthonomus rubi*, Strawberry root weevil *Otiorhynchus rugostriatus*, Vine weevil *Otiorhynchus sulcatus* (see Weevils). May attack young and established plants. **Symptoms:** adult weevils feed on the leaves, biting pieces out of them; larvae feed on the roots, causing wilting. **Control:** spray in the third week of April to kill weevils feeding on the leaves, using Carbaryl; spray in early September with Carbaryl for red-legged weevils; larvae can be controlled by drenching the plants with Carbaryl in September. **30** Wireworms *Agriotes* species (see Wireworms). May attack at all stages. **Symptoms:** wilting and possibly death of plants due to wireworms feeding on the roots. **Control:** use gamma-HCH dust.

Sun Green plants must have light in order to grow, but excess sun can literally scorch or burn many plants, for example, large leaved begonias, which in natural conditions would grow only in shade. Many greenhouse plants at propagating require shading.

Sunflower *Helianthus annuus* and varieties. Hardy annual. Height: tall. Flowering time: summer to late summer. Flower colours: orange, reddish-yellow, yellow. Use: borders, preferably against walls or fences.

Pests and Diseases. 1 Grey mould, caused by the fungus *Botrytis cinerea* (see Botrytis). May attack young and established plants. **Symptoms:** rotted leaves covered with white mould. **Control:** spray with Benomyl. Attack is usually on plants with soft growth, so in future use more potash in preparing ground. **2** Sclerotinia rot, caused by the fungus *Sclerotinia sclerotiorum*. **Symptoms:** brown spots and white mould covering stems and flower heads; black sclerotia (resting bodies of the fungus) on the spots. Often associated with Grey mould (see 1 above Grey mould). **Control:** spray with Benomyl. **3** Slugs. May attack young plants (see Slugs and snails).

Surface caterpillars   (see Cutworms)

Swede   Derived from a cross between a turnip variety and another brassica (see Turnip, Brassicas). Grown mainly for cattle feeding. For **Pests and Diseases**, see Brassicas.

Sweet pea *Lathyrus odoratus* Sweet pea varieties, *Lathyrus latifolius* Everlasting pea and varieties and other *Lathyrus* species. Mostly hardy annuals and herbaceous perennials. Height: climbing, medium. Flowering time: in the greenhouse, spring; outdoors, summer. Flower colours: cream, blue, lilac, purple, red, white, or parti-coloured. Very fragrant. Use: in the open, trained to supports, fences or walls, commercial or garden cut flowers, annual borders, herbaceous borders, summer bedding, containers, window boxes, in borders or pots in the greenhouse, commercial glasshouse cut flowers.

Pests and Diseases. 1 Bud drop (see Physiological disorders). May affect established plants. **Symptoms:** flower buds drop before opening. In the greenhouse and possibly in the open also, variation in temperature is the main cause. Sweet peas react slowly to sudden or sharp temperature changes and flower buds may drop because of this. Temporary shortage of water may be another cause. **Control:** mulch with suitable organic matter to conserve the moisture in the soil; a sheltered site against a sunny wall often helps to avoid drastic drops in night temperature. **2** Downy mildew (see Mildew, Pea 5 Downy mildew). **3** Foot rot (see Pea 8 Fusarium foot rot or Fusarium wilt). **4** Leafy gall *Corynebacterium fascians* (see Bacterial diseases). May attack young and established plants. **Symptoms:** dense leafy growth develops from ground level. **Control:** maintain standards of cleanliness in all details when raising young plants; practise long rotation with peas and sweet peas (see Rotation). **5** Mildew (see Mildew, Pea 5 Downy mildew). **6** Powdery mildew (see Pea 19 Powdery mildew). **7** Virus (see Virus diseases). (a) Mosaic virus. **Symptoms:** leaves paler green between the veins and distorted; plants tend to be less vigorous than normal. This virus is spread by aphids. **Control:** spray regularly to control aphids. Plants may appear to grow out of disease, especially if fed with extra nitrogen, since strong green growth tends to mask the typical light and dark green markings caused by virus infection. (b) Streak, caused by the streak virus. **Symptoms:** dark streaks on the leaves, leaf stalk and stems. This virus is also spread by aphids. **Control:** spray regularly with Dimethoate, Fenitrothion or

Malathion to control aphids. As with mosaic, plants may appear to grow out of this disease, especially if fed with extra nitrogen. However, over-feeding with nitrogen may induce streak virus on other plants, such as tomatoes, and on other sweet peas too. **8** White mould, caused by the fungus *Cladosporium album.* May attack at all stages. **Symptoms:** similar to those of powdery mildew (see 6 above Powdery mildew); in severe attacks, leaves decay and fall. **Control:** Spray with Benomyl or Zineb; avoid growing sweet peas in damp humid conditions.

Sycamore (see Maple)

Symphilids Symphilids are small, white, wingless insects, with 12 pairs of legs. They move actively, and prefer moist soil conditions. Young and mature insects feed on the young roots of plants, causing blue discolouration and wilting, and encouraging infection by diseases.

Control: lift and examine a few affected plants and if there are 10 or more symphilids at the roots, drench the soil with Diazinon or gamma-HCH.

Syringa (see Lilac)

Systemic chemicals Systemic pesticides are absorbed into the sap flow of plants and are then distributed throughout the parts above ground. Systemic fungicides are available as are systemic insecticides (see Appendix p. 245–54). Because of their systemic action, it is not absolutely necessary to obtain perfect cover of a plant with a spray, and in some cases, a root drench will control foliar pests and diseases.

T

Tagetes (see Marigold)

Thrips Thrips are commonly known as thunderflies because of their habit of swarming in hot, humid weather. They are minute, thin insects, usually black, with narrow feathery wings, the immature stages being fatter and yellow. Damage is easily recognized as silvery speckling of leaves and flowers. The discoloured areas have small black or red specks of waste material, although the mere presence of adults in large numbers is often enough to disfigure flowers. Important species are gladiolus thrips *Thrips simplex* which attack gladioli and freesias, glasshouse thrips

Thrips, commonly known as thunderflies

Heliothrips haemorrhoidalis, a pale-coloured species, which is found on many crops under glass, especially cucumbers and azaleas, and onion thrips *Thrips tabaci* which is most troublesome on carnations.
Control: With the exception of these examples, it is rarely necessary to take special steps to control thrips because they are readily eliminated by pesticide treatments for more important pests; a single spray of gamma-HCH or any other contact pesticide usually proves adequate on outdoor crops, fumigation being most effective under glass.

Thuja *Thuja occidentalis* and varieties, and *Thuja orientalis* and varieties. Hardy, evergreen and coniferous trees and shrubs. Height: large trees, small to medium shrubs. Foliage: decorative with colours varying from clear and dark green to bronze and golden. Growth: attractive and varied. Use: trees planted singly or in groups or avenues, shrubs in borders or rock gardens.
Pests and Diseases. 1 Leaf scorch, caused by the fungus *Didymascella thujina*. May attack established plants. **Symptoms:** browning of leaves at the base of trees; tiny brown cushions on the upper sides of the leaves. **Control:** spray with Bordeaux mixture as soon as the disease has been seen and confirmed. **2** Root rot, caused by the fungi *Armillaria mellea, Fomes annosus* (see Armillaria). **3** Rust, caused by a *Melamposorella* species of fungus (see Rust).

Tilia (see Lime)

Tissue The term for the fabric of organic structures which consists of an aggregate of similar cells.

Tomato Tomato varieties are complex hybrids derived from

Lycopersicon esculentum (see Hybrid). Greenhouse (half-hardy in mild areas only) herbaceous perennials. Height: small to tall. Flowering time: spring, summer. Flower colour: yellow. Fruit colours: pink, purple-red, red or yellow. Foliage: hairy and strong smelling. Use: food crop, grown for its fruit, borders or containers in greenhouse, commercial glasshouse and other structures, containers or borders outdoors in good summer areas only. The plants are naturally sprawling in habit, rooting at stem joints, and are trained to supports, but there are also compact bushy varieties.

Pests and Diseases. 1 Aphis, *Aulacorthum solani*, *Myzus persicae* and other species (see Aphis). May attack at all stages. **Symptoms:** aphids on the plants; honeydew on plants and fruit (see Honeydew); yellowing of the foliage due to removal of sap by aphids feeding. It is important to control aphids, not only for the damage they do, but also because they can transmit certain virus diseases from infected to healthy plants (see Virus diseases). **Control:** there is a wide choice of suitable insecticides; for young plants, use sprays with Oxydemeton-methyl, Dimethoate or Malathion, repeating the spray in 10 days; for established plants, use gamma-HCH or Nicotine smokes; sprays of organophosphorus compounds which have systemic action may be used, but aphids may develop a resistance to some of these compounds (see Systemic). **2** Bacterial canker *Corynebacterium michiganense* (see Bacterial diseases). May attack established plants. **Symptoms:** small, light brown areas on the surface of the leaves which coalesce and kill the leaves; raised yellowish-white mealy areas on the stems; white spots with dark centres known as bird's-eye spots on the fruit; the skin of the stems separates easily from the woody centre. **Control:** reduce temperature by ventilation and cease overhead watering; remove infected plants carefully and spray remaining plants with a mild copper fungicide every 3 days until the disease is under control, then spray weekly; general hygiene and soil sterilization must be carefully observed (see Soil sterilization). General hygiene includes the routines of removing and destroying all remains of a finished tomato crop, including roots, strings and all debris, washing down the greenhouse with detergent or disinfectant, fumigating with Formalin or sulphur (do not use sulphur in metal greenhouses), and sterilizing canes and containers. **3** Blight, caused by the fungus *Phytophthora infestans*. May attack established plants. Not common in the greenhouse. **Symptoms:** brown areas on the leaves; dark streaks on the stems; reddish-brown marbled patches on green fruit, which later shrivel and become useless. **Control:** reduce humidity in the greenhouse by ventilating and giving a little heat; spray regularly with fungicides suitable for potato blight, such as Maneb, Mancozeb (commercially) or Zineb, especially if blight is present on potatoes nearby (see Potato 6 Blight). **4** Blossom end rot (see Physiological disorders). May occur on established plants. **Symptoms:** dark green circular patches on the blossom end or bottom of fruit, later becoming sunken and black. Lack of water in the soil or high concentrations of fertilizer salts in the soil, especially if the soil is short of lime, may be the cause. **Control:** water carefully and regularly; avoid over-feeding and check lime content. **5** Blotchy ripening (see Physiological disorders). May occur on established plants. **Symptoms:** pale patches which do not colour on the fruit. This is due to irregular feeding and watering, and is most obvious in strong-growing varieties. **Control:** ensure regular feeding and watering and maintain even temperatures as far as possible; avoid flooding with plain water, which may upset the balance of nutrients in the soil. **6** Boss fruit (see Physiological disorders). May occur on established plants. **Symptoms:** hollow fruits. This disorder is due to temperature variation, especially excess heat, caused by weather conditions. **Control:** regulate temperatures in the greenhouse as far as possible. This trouble is usually only temporary. **7** Bronzing (see Physiological disorders). May occur on established plants. **Symptoms:** the surface of the fruit has a dull metallic appearance, due to a layer of dead cells immediately below the skin. There are various causes, such as excessively high daytime temperatures, virus infection (see 42 below) or Boron deficiency (see Physiological disorders). This trouble does not usually affect more than a few trusses of fruit. **Control:** avoid high daytime temperatures, and check other possible causes. **8** Brown rot and corky root, caused by the fungus *Pyrenochaeta lycopersici*. May attack established plants. **Symptoms:** plants become unthrifty, and wilt in bright weather; later, larger roots show brown uneven cork-like swellings; stem bases rot; lower leaves turn yellow and wither. **Control:** use a stem base drench of Nabam (commercially only) after planting to reduce infection; carefully observe general hygiene (see 2 above) and practise soil sterilization (see Soil sterilization). **9** Buckeye rot, caused by the fungus *Phytophthora parasitica* and other

Phytophthora species. May attack established plants. **Symptoms:** grey to reddish-brown zoned patches on the fruit. **Control:** support lower trusses to keep them clear of the ground, and so prevent soil, which carries infection, from coming in contact with the fruit; spray the surrounding ground and the lower trusses with a proprietary copper fungicide to prevent the disease from spreading. **10** Caterpillars (see 40 below Tomato moth). **11** Damping off, foot rot and root rot, caused by fungi of *Pythium*, *Phytophthora* and *Rhizoctonia* species (see Damping off, Foot rot, Root rot, Stem rot). May attack seedlings and young plants. **Symptoms:** seedlings or young plants collapse at soil level; roots may be rotted. **Control:** drench soil surface with Cheshunt compound, Zineb or a proprietary copper compound; always sterilize seed containers and compost; sterilize border soil; incorporate Quintozene (commercially) in composts if Rhizoctonia is suspected; ensure a clean water supply; do not plant in soil with a temperature less than 13°C (56°F). **12** Dry set (see Physiological disorders). May occur on established plants. **Symptoms:** flowers drop off or fruit fails to set. **Control:** always damp down during the morning on sunny days as pollination will fail if the atmosphere is too dry; use proprietary hormone setting compounds (see Hormone), but avoid excessive application as distorted fruits may occur on lower trusses. **13** Flower abortion (see Physiological disorders). May occur on established plants. **Symptoms:** a large proportion of small fruits, known as chats, develop which are of little use; although the flowers are fertilized, fruits do not develop. This is thought to be due to the production of poor pollen, resulting from insufficient light, or from an unsuitable balance between day and night temperatures. **Control:** it is very important to grow tomato plants as well as possible from start to finish as the flower trusses are actually initiated at a very early stage in the life of the plant, and adequate light during this process is essential; during the propagation period, take great care to ensure suitable conditions of light and temperature, and also of ventilation, watering and atmospheric humidity. In poor light areas, commercial growers make use of supplementary lighting, or of growing rooms. **14** Flower drop (see Physiological disorders). May occur on established plants. **Symptoms:** flower stems break and drop off. May be due to lack of humidity, lack of moisture at the roots, or a high concentration of fertilizer salts on the soil. **Control:** correct growing conditions by adjusting watering, humidity and feeding. **15** Flowers missing (see Physiological disorders). May occur on established plants. **Symptoms:** flowers form on the truss, but open only partially, or flowers may be missing. The cause may be too much nitrogenous feeding, or temperatures too low at night and too high by day. **Control:** apply potash to harden growth and adjust feeding; regulate day and night temperatures. **16** Fruit splitting (see Physiological disorders). May occur on established plants. **Symptoms:** fruit cracks or splits. This trouble is more common in cold greenhouses where fruit development has taken a long time and the skin has grown tough. It is caused by irregular water uptake, and sometimes by varying temperatures. **Control:** water regularly and carefully; keep temperatures as even as possible; use shading when necessary, to prevent excessive daytime temperatures. **17** Ghost spots (see **23** below Grey mould). **18** Glasshouse red spider mite *Tetranychus urticae* (see Mites). May attack at all stages. The mite has a small pear-shaped body, yellow-green or red in colour. **Symptoms:** foliage hard and parchment-like, with yellow mottling on the upper surface changing to complete yellowing; webs are produced on plants and all stages of the pest are visible; fruit and flowers are spoilt by webbing. **Control:** use one of the many acaricides which are available (see Acaricides); for seedlings, spray with Dimethoate, Oxydemeton-methyl or Malathion and for established plants, use Dicofol in addition to these; the mites may develop a resistance to a particular chemical as aphids do, so vary the materials used; as the mites hibernate during the winter, come out of hibernation in spring and breed rapidly, it is important to remove old plants immediately the fruit is finished to prevent the mites from hibernating; biological control of mites, using predators, may be effective (see Biological control). **19** Glasshouse symphilid *Scutigerella immaculata*. May attack at all stages. Adults have white bodies and 12 pairs of legs, and move rapidly. They prefer moist to dry soil. All stages of this pest feed on young roots. **Symptoms:** wilting and blue discolouration of the plants, probably followed by infection by various diseases. When 10 or more symphilids are found at one root system, control is necessary. **Control:** drench soil with Diazinon or gamma-HCH. **20** Glasshouse whitefly *Trialeurodes vaporariorum* (see Whitefly). May attack at all stages. **Symptoms:** whitefly gather in large numbers on young plants and on the growing points of established plants; the small adults, with white wings, rise in clouds when foliage is disturbed; the scale-like larvae can be seen on the plants; dead areas on the leaves, caused by whitefly removing the sap; leaves are coated with honeydew (see Honeydew). **Control:** spray with Diazinon or Malathion; biological control is possible but not always practical (see Biological control). **21** Golden potato cyst eelworm *Globodera rostochiensis* (see Eelworms). May attack established plants. **Symptoms:** plants are dwarfed, tinged with a purple colour and tend to wilt; large numbers of fibrous roots in evidence near the surface of the soil; if, from mid-summer onwards, plants are lifted and the roots examined, white or golden pinhead-sized cysts are visible on the roots (see Eelworms). All other stages in the life-cycle are microscopic. **Control:** help plants to form extra roots by drawing soil up over the old root system and giving extra water; strict hygiene is necessary, as this pest is very easily spread on soil, tools and containers. (Potatoes, and the ground they grow in or have grown in, may be a source of infection.) **22** Greenback (see Physiological disorders). May attack established plants. **Symptoms:** the tops of the fruit stay green and hard. Too hard defoliation, lack of potash or excess sunlight may cause greenback. **Control:** increase applications of potash, shade from bright sunshine in very hot weather and do not over defoliate; if greenback is troublesome, grow varieties such as Moneymaker and varieties similar to it which have been bred to be free of greenback; avoid Ailsa Craig and varieties bred from it which are susceptible to greenback. **23** Grey mould, caused by the fungus *Botrytis cinerea* (see Botrytis). May attack young and established plants. **Symptoms:** pale brown marks on the stems, with, in humid conditions, grey mould growing on them; fruit show green rings with pinpoint brown centres, which are called ghost spots, and soft, mouldy rotting. **Control:** as for stem rot (see 36 below Stem rot); avoid any checks to plant growth; avoid conditions of high humidity by ventilating and, if necessary, use a little heat to create a buoyant atmosphere; after de-leafing, spray with Benomyl or Dichlofluanid, or use Tecnazene smokes. **24** Iron deficiency (see Physiological disorders). May occur at all stages. Younger growth is affected first. **Symptoms:** general yellowish blanching of the leaves, with the main veins remaining green. **Control:** apply a proprietary chelated iron compound to the soil round the affected plants (see Sequestrols). This trouble is most common when the soil is alkaline, so adjust the pH to less than 7 before growing tomatoes. **25** Leaf curling (see Physiological disorders). May occur on established plants. **Symptoms:** leaves, especially older leaves, curl upwards excessively. This is due to great variation in day and night temperatures, which makes it difficult for the plants to deal with their stored food material. **Control:** avoid, as far as possible, very marked differences in day and night temperatures. **26** Leaf mould, caused by the fungus *Cladosporium fulvum*. May attack established plants. **Symptoms:** yellow patches on the upper surfaces of the leaves, with corresponding brown or purplish, velvety spots on the lower surfaces. **Control:** avoid conditions of high humidity by ventilating and giving a little heat to create a buoyant atmosphere; always allow reasonable space between plants; use resistant varieties. **27** Leaf scorch (see Physiological disorders). May occur on young plants, especially at the propagating stage or after planting out, but also on established plants. **Symptoms:** leaves turn white or brown, usually in patches. This is caused by excess heat and insufficient watering or possibly by industrial fumes or paraffin. However, it is not usually damaging in the long term, as plants grow out of it. **Control:** adjust growing conditions and check on fumes. **28** Magnesium deficiency (see Physiological disorders). May occur at all stages. **Symptoms:** yellowing and browning between the veins on lower leaves which later curl up and die. The condition progresses up the plant. **Control:** spray with a solution of magnesium salts (Epsom salts) at 20 g per litre (3 oz per gallon) and repeat at intervals; reduce potash fertilizer applications temporarily, as potash aggravates the condition. **29** Manganese deficiency. **Symptoms:** mottling of leaves on young plants. This condition is generally transient. **30** Manganese toxicity (see Mineral deficiencies). May occur at all stages. **Symptoms:** brownish marks on stems and leaf stalks; leaves may droop and wither. It occurs most frequently in acid soils, particularly after soil sterilization. **Control:** check the soil pH before growing tomatoes and if it is too low, adjust the pH to 6–6.5 by adding a suitable form of lime. **31** Oedema (see Physiological disorders). May occur in established plants. **Symptoms:** blisters or blotches on stems or leaves. These form when the plants are unable to give off water through the leaves rapidly enough, due to excess humidity in the greenhouse, or because oil-containing sprays have been used. The condition is more common in container-grown plants if excess water is left lying between the containers. **Control:** lower humidity by avoiding excess water; ensure adequate ventilation, especially at night. **32** Red spider (see 18 above Glasshouse red spider mite). **33** Rootknot eelworm *Meloidogyne* species (see Eelworms). May attack established plants. **Symptoms:** lower leaves pale and severely wilted; large, irregularly-shaped galls on the roots, which distinguish this pest from the golden potato cyst eelworm (see 21 above Golden potato cyst eelworm). **Control:** if the soil is infested, remove damaged plants and roots and steam sterilize or use D-D injection (commercially) but if this is impossible, use pot or peat culture methods, using sterilized materials; always buy healthy plants from a reliable source; commercial growers may use plants grafted on to rootstocks which are resistant to some species of eelworm. **34** Silvering. Thought to be a genetical disorder, that is, it is a change occurring in the plant's make-up. **Symptoms:** foliage turns light in colour on part of the plant (up to half the leaves may be affected). The silvery appearance is due to a variation in the layers of tissue in the leaves. **Control:** no control is possible. The condition may correct itself or it may persist. **35** Springtails *Collembola* species. May attack seedlings and young plants. **Symptoms:** pinholes or scraping of the surface of the foliage evident in seedlings and young plants; if some soil from around the base of the stem is floated in a bucket of water, large numbers of the minute, white or colourless and wingless springtails will be seen. **Control:** apply gamma-HCH dust to the soil, or drench the soil with gamma-HCH or Malathion. **36** Stem rot, caused by the fungus *Didymella lycopersici* (see Stem rot). May attack established plants. **Symptoms:** brown, grey or blackish rot of base, and later of upper parts, of stems; raised, dark spots, which are the fruiting bodies of the fungus, may be present on the rotting areas; roots rotted; dark areas on the lower, flower end of the fruits. **Control:** always remove and destroy all remains of a finished tomato crop, including roots, strings and all debris in accordance with general hygiene; wash down the greenhouse with detergent or disinfectant; fumigate with Formalin or sulphur, but do not use sulphur in metal greenhouse; sterilize canes and containers; sterilize soil before planting tomatoes again; after planting, spray Captan on stem bases and adjacent ground, and repeat 3 weeks later or spray stem bases with Maneb (combined form), and repeat 3 weeks later, but do not apply Maneb to the ground, or again after planting, drench stem bases with Benomyl at the rate of 750 ml (1 pint) per plant, and later spray plants with Benomyl in the usual way; always inspect plants regularly during cropping, and remove immediately any that show signs of infection. **37** Symphilids (see 19 above Glasshouse symphilid). **38** Thrips *Thrips tabaci* and occasionally other *Thrips* species. May attack established plants. **Symptoms:** wilting and scarring of foliage and fruit caused by thrips feeding on sap; minute, yellow-brown thrips with strap-like wings may be visible. **Control:** spray foliage with Malathion, giving 2 applications at 14-day intervals. Thrips are also important as they spread spotted wilt virus (see 42(c) below Spotted wilt virus). **39** Tomato leaf miner *Liriomyza bryoniae*. May attack at all stages. Particularly damaging to seedlings, which may be killed. **Symptoms:** adults feed on foliage, where they cause pitting; small, pale orange larvae tunnel in the seed leaves of seedlings, often killing the seedlings. **Control:** spray soil and seedlings with Diazinon, Dimethoate or Malathion. This pest is chiefly confined to the south of England. **40** Tomato moth *Lacanobia oleracerea*. May attack young and established plants. **Symptoms:** caterpillar has a pale green head, and yellow-brown body with dark stripes on the back and yellow stripes on the sides. During the summer, young caterpillars skeletonize the leaves, older caterpillars strip the leaves and eat holes in the fruits and stems. Adult moths are purple-brown in colour. Caterpillars of other moth species may attack in the same way. **Control:** hand-pick caterpillars from young plants; spray established plants with Dichlorvos or Trichlorphon at 1-week intervals. **41** Verticillium wilt (see 44 below Wilts). **42** Virus (see Virus diseases). (a) Aspermy, Tomato aspermy virus. **Symptoms:** plants appear bushy, with distorted and mottled foliage; fruits may be small and virtually without seeds. **Control:** control aphids (see 1 above Aphis) which can transmit the disease from neighbouring tomatoes, or from chrysanthemums, which can also harbour the disease. (b) Cucumber mosaic virus, and (c) Spotted wilt virus. **Symptoms:** foliage shows mottling, bronzing, ringspots and distortion; plants may be stunted or killed. **Control:** control aphids (see Aphis) and thrips (see Thrips) which transmit some of the virus diseases. (d) Streak, (e) Tobacco mosaic virus and (f) Tomato mosaic virus. **Symptoms:** seedlings turn purple until growth is checked; on older plants, mosaic viruses cause light and dark green mottling of the leaves, with variable distortion; leaf blades may not form, giving a fern-like effect; streak virus causes grey streaks on stem and leaf stalks, and bronzing and sunken markings on the fruit. **Control:** use virus-treated and tested seed; follow hygiene as for stem rot (see 36 above Stem rot), except that washing down should be done with 3% trisodium phosphate; fumigate with Formalin or sulphur but avoid using sulphur in metal greenhouses; sterilization with steam will reduce virus in the soil but chemicals are ineffective; during growth, handle obviously infected plants last when de-leafing and side-shooting; wash hands before handling plants; maintain well-balanced feeding of plants to minimize effect of virus; if fruit bronzing is persistently bad, grow resistant varieties. **43** Whitefly (see 20 above Glasshouse whitefly).

44 Wilts, caused by the fungi *Fusarium redolens, Fusarium oxysporum* fusarium species *lycopersici, Verticillium albo-atrum* and *Verticillium dahliae.* May attack young and established plants and may affect one side only especially in the case of *Fusarium.* **Symptoms:** yellowing and/or wilting of the leaves progressively up the plant; woody stems often discoloured inside, becoming light, dark or reddish-brown; severely affected plants may die. **Control:** follow hygiene and soil sterilization as for stem rot (see 36 above Stem rot); drench the stem bases after planting with Benomyl, at the rate of 750 ml (1 pint) per plant, to reduce infection; grow resistant varieties (see Resistant varieties) or use plants grafted on to resistant rootstocks. **45** Woodlice, Slaters *Armadillidium* species. May attack at all stages. Particularly damaging at the seedling stage. **Symptoms:** stems of seedlings chewed through at or below soil level; roots stunted. Woodlice usually hide under rubbish, boxes and pots during the day, coming out at night to feed. The body is grey-brown, segmented and armour-plated, and there are numerous small legs on the underside. **Control:** remove all rubbish, and rubbish boxes; apply gamma-HCH spray or dust, or Carbaryl dust, to the soil.

Trace elements   In plant nutrition, certain major elements, such as nitrogen, phosphorus and potassium, are essential to normal growth and are taken up as compounds dissolved in the soil water. These are known as trace elements. The trace elements are essential to normal growth, but only in extremely minute amounts. In excess, they may cause damage. Other elements are equally important but are acquired in other ways.

Tulip   Tulip varieties are complex hybrids (see Hybrid), far removed from the *Tulipa* species from which they are derived. A few *Tulipa* species are grown, chiefly in rock gardens. Tulip varieties are divided into a number of classes, according to type and flowering time. Hardy herbaceous perennials, with bulbs. Height: dwarf to medium. Flowering time: varies with method of growing from spring to early summer outdoors, winter to spring in the greenhouse or indoors. Flower colours: blue-black, bronze, cream, green, lilac, orange, pink, purple, red-purple, red, white, yellow or parti-coloured. Use: spring bedding, borders, rock gardens, garden and commercial cut flowers, outdoor containers, window boxes, greenhouse and indoor pot plants, commercial glasshouse cut flowers and pot plants.

Pests and Diseases. 1 Aphis, Tulip bulb aphid *Dysaphis tulipae* (see Aphis, Lily 1 Aphis). **2** Bulb mite (see Bulb mite). **3** Chalkiness (see Physiological disorders). **Symptoms:** hard and chalky areas on the inner scales of bulbs. **Control:** avoid damage during handling; avoid exposure to hot sunshine; store under the best possible conditions. Chalkiness may be associated with storage rots and bulb mites (see Bulb mite), though the condition is primarily due to mechanical damage and unsuitable conditions. **4** Eelworm (see 13 below Stem and bud eelworm). **5** Fire, caused by the fungus *Botrytis tulipae*. May attack at all stages. A serious disease. **Symptoms:** on bulbs, sunken circular spots with raised margins under the outer scales; black sclerotia (resting bodies of the fungus) on these spots; on growing plants, dying shoots; spots which later develop grey mould, on flowers and leaves; in a severe attack, failure of flower buds to open. **Control:** after lifting, discard obviously infected bulbs; soak the apparently sound bulbs in Benomyl, or dust with Quintozene commercially; before planting, incorporate Quintozene dust in the surface soil; sterilize border soil, and soil used for container and pot bulbs (see Soil sterilization); as soon as infection is seen in growing plants, remove and destroy them; spray remaining plants with Captan, or Zineb, Maneb or Mancozeb/Zineb commercially. **6** Grey bulb rot, caused by the fungus *Sclerotium tuliparum*. **Symptoms:** failure of shoots to emerge, or shrivelling and dying of shoots shortly after emerging; soil sticking to the noses of the bulbs; spreading of rot from nose downwards until bulbs are covered with fungal growth; bulbs embedded with small black or brown resting bodies of the fungus. **Control:** destroy diseased bulbs. Before planting tulips again, thoroughly

Tulip 7 Hard scale

sterilize the border soil, or incorporate Quintozene dust in the surface soil commercially and use sterilized soil for pots and bowls (see Soil sterilization). **7** Hard scale (see Physiological disorders). **Symptoms:** hard outer skins of bulbs preventing roots from emerging and forcing roots upwards between outer skin and fleshy scales; poor growth; failure of flowers to appear. **Control:** if this condition is noticed, cut outer skin carefully round base of bulbs before planting, to allow roots to emerge. **8** Mice (see Mice). **9** Penicillium bulb rot, caused by a *Penicillium* species of fungus. **Symptoms:** blue-green mould on bulbs. **Control:** not necessary if bulbs are grown in good conditions. **10** Root rot, caused by *Pythium* species of fungi. **Symptoms:** rotting of roots at tips. **Control:** not necessary if bulbs are grown in good conditions. **11** Shanking, caused by *Phytophthora* species of fungi. **Symptoms:** failure to emerge or drooping and dying of flower stalk; failure of flowers to open; poor growth; brownish rot up the centres of bulbs. **Control:** ensure that compost and containers are thoroughly sterilized by steaming, or by using Formalin. **12** Slugs (see Slugs and snails). *Note*: incorporate slug baits containing Methiocarb into the soil surrounding bulbs outside in spring. **13** Stem and bulb eelworm *Ditylenchus dipsaci* (see Eelworms). A serious pest. **Symptoms:** lopsided leaves bearing flaky scars on growing plants in spring; twisted, puffy and horribly distorted stems; green petals on flowers; silvery and soft bulbs; whitish woolly substance exuding from the necks of bulbs. **Control:** destroy affected bulbs; buy good quality bulbs and plant on fresh ground. Commercial growers use hot water treatment, soaking bulbs for 3 hours at 44°C (112°F), (see Hot water treatment) but this is not suitable for tulips intended for forcing, or bulbs are dipped in a solution of Thionazin (0.23%). **14** Topple (see Physiological disorders). **Symptoms:** soft areas on stems as flower buds are developing, causing stems and flowers to collapse. May be associated with conditions during raising and storage of the bulbs. **Control:** with indoor bulbs, take care not to bring bowls of bulbs into warm rooms too early, or keep them at too high a temperature. Bulb forcers should take care not to force too soon or at too high a temperature. A preventative spray with calcium nitrate may be applied to varieties known to suffer particularly from topple. **15** Virus (see Virus diseases). (a) Breaking, caused by various virus diseases. **Symptoms:** marked colour patterns on flowers; little effect on leaves. **Control:** in gardens remove plants which have patterned flowers and spray with insecticide to control aphids which spread the disease. Commercial growers remove infected plants as soon as patterned flowers appear and keep stocks of bulbs that have shown infection separate from healthy bulbs, as well as controlling aphids. The patterned flowers are very attractive and in gardens, no effort need be made to get rid of them. At one time they were highly valued and a few varieties are still grown and sold. (b) Streak *Tobacco necrosis* virus. **Symptoms:** distorted leaves, with brown streaks which may coalesce forming larger withered areas. **Control:** remove and destroy diseased plants; sterilize soil to kill the fungus which, in this case, transmits the virus (see Soil sterilization).

Turf (see Lawns)

Turnip Varieties derived from *Brassica rapa*. Hardy biennial, with swollen, fleshy root. Height: small to medium. Roots: long or round, and fleshy, brownish-green or greenish-yellow. Use: food crop, grown for its roots. For **Pests and Diseases**, see Brassicas.

Turnip fly (see Flea beetles)

U

Ulmus (see Elm)

V

Variety Botanically, a variety is a variation of a species (see Species) occurring in the wild. For example, the marsh marigold *Caltha palustris* has single golden flowers, but there is a double-flowered variety *Caltha palustris plena*. In horticulture, the word 'variety' has long been used for any variation of a species or hybrid (see Hybrid) occurring naturally or produced by deliberate breeding or in any other way. Some time ago an effort was made to introduce the term 'cultivar' (see Cultivar) for cultivated varieties, but this has not come into general use. Throughout this book the term 'variety' is used for cultivated varieties.

Vegetable marrow Varieties and hybrids of *Cucurbita pepo ovifera* (see Hybrid). Half-hardy and greenhouse annual. Height: trailing. Flowering time: throughout summer. Flower colour: yellow. Fruit colour: green, white, yellow or parti-coloured. Use: food crop grown for the fruit; may be grown in the greenhouse, trained up posts, or grown in frames.

Pests and Diseases. 1 Glasshouse millipede *Oxidus gracilis* (see Cucumber 13 Millipedes). **2** Grey mould, caused by the fungus *Botrytis cinerea* (see Cucumber 11 Grey mould). **3** Millipede (see 1 above Glasshouse millipede). **4** Powdery mildew, caused by the fungus *Erysiphe cichoracearum* (see Cucumber 14 Powdery mildew). **5** Slugs and snails (see Slugs and snails). **6** Virus diseases, Cucumber mosaic virus. **Symptoms:** pale yellow lines or rings on the leaves; considerable puckering; shortened stems with leaf joints closer together than normal; fruits mottled with wart-like areas on them. **Control:** (see Cucumber 20 Virus diseases).

Veronica (see Hebe) *Veronica longifolia*, *Veronica spicata*, *Veronica teucrium*, *Veronica prostrata*, *Veronica virginica*, and all their varieties as well as other Veronica species and varieties. Hardy herbaceous perennials. Height: creeping to large. Flowering time: summer. Flower colours: cream, blue, pink, purple or white. Use: herbaceous borders, rock gardens.

Pests and Diseases. 1 Downy mildew, caused by the fungus *Peronospora grisea* (see Mildew). **2** Leaf spot, caused by the fungus *Septoria veronicae* and other fungi (see Leaf spot). **3** Powdery mildew, caused by the fungus *Sphaerotheca humuli* (see Mildew).

Vine (see Grape vine)

Viola The genus *Viola* includes species from which the garden pansy, the sweet violet and other violets, and the viola, with all their varieties, have been derived. Some *Viola* species may also be grown in gardens. Hardy, chiefly herbaceous perennials. A few annuals, biennials and shrubs. Height: dwarf to small. Flowering time: spring to summer. Flower colours: cream, blue, brown, grey, lilac,

orange, purple, white, yellow or parti-coloured. Use: in the open, beds, borders, edging plants, outside containers, window boxes, commercial cut flowers.
Pests and Diseases. 1 Aphis, Violet aphid *Myzus ornatus.* **Symptoms:** acute distortion of leaves. **Control:** spray with insecticides to control aphids which not only do damage, but also spread virus diseases. **2** Cutworms (see Cutworms). **3** Foot rot, caused by various species of fungi. May attack established plants. **Symptoms:** wilting; dull green petals. **Control:** plant new plants on fresh ground, if this disease has been troublesome; treat soil with Quintozene before planting; lime the soil. In extreme cases, sterilize the soil with Basamid (Dazomet) (see Soil sterilization). **4** Leaf spot, caused by various species of fungi (see Leaf spot). **5** Rust, caused by a *Puccinia* species of fungus (see Rust). May attack established plants. **Symptoms:** pale green spots on leaves in spring; light brown pustules in summer. **Control:** spray with Mancozeb, Maneb or Zineb when first seen. **6** Slugs (see Slugs and snails). **7** Smut, caused by the fungus *Urocystis violae.* May attack established plants. **Symptoms:** dark purple swelling on leaves and leaf stalks. **Control:** remove and destroy diseased plants as soon as possible; plant new plants on different ground only. **8** Stem rot, caused by various species of fungi. May attack established plants. **Symptoms:** rotting stems near base and wilting plants. **Control:** plant new plants on fresh ground. **9** Virus colour breaking caused by various virus diseases (see Virus diseases). **Symptoms:** mottling of leaves and variation of colour in flowers. **Control:** spray with insecticides to control insects which spread the diseases; buy new plants from a reliable source.

Virginia creeper *Parthenocissus quinquefolia,* other *Parthenocissus* species including *Parthenocissus tricuspidata veitchii,* which is often called *Ampelopsis veitchii.* Hardy deciduous shrubs. Height: large and climbing. Foliage: ornamental, vivid reds and oranges in autumn. Fruit: black or blue. Use: vigorous climbers for fences, walls and other supports.
Pests and Diseases. 1 Downy mildew, caused by the fungus *Plasmopara viticola.* May attack established plants (see Mildew). **2** Leaf spot, caused by weather and by various species of fungi. **Symptoms:** marking and spotting of leaves. A certain amount of spotting is usual but is seldom destructive. **Control:** spray with fungicide, though this is seldom possible on well-developed

specimens. **3** Wilt caused by a *Cladosporium* species of fungus (see Wilt). May attack established plants. **Symptoms:** wilting of leaves. **Control:** effective control difficult, as disease is internal in plant tissue.

Virus diseases There are many more plant virus diseases than is generally appreciated. The word 'virus' means poison. Virus diseases live within the cells of the host plant (see Host) as ultra-microscopic rod-shaped organisms, and compete with the plant for vital food supplies. Some plants are tolerant to a certain degree, or may have resistance to virus infection. Others are not tolerant or resistant, and leaves, flowers, fruit, or whole plants may show discolouration, mottling, distortion, stunting or any of a wide range of symptoms, and may even die. Virus diseases may be rapidly spread in a number of ways. For example, they may be carried on the hands, on tools, on seed or by pests which suck sap from infected plants, then spread infection when they go on to feed on healthy plants. It may be almost impossible to control virus diseases once plants are infected, although there have been recent successes by inoculation methods, whereby a weak strain of virus releases antibodies and antigens which resist further attacks by stronger forms of virus. An example is the inoculation of tomato against tomato mosaic virus. A wide range of methods may be used to clean stock of virus, including meristem culture (see Meristem culture), cell culture (see Meristem culture) and heat treatments (see Heat treatments). **Control:** for amateur gardeners: lift and burn virus-infected plants when symptoms are noticeable; control the various sucking pests which may spread virus diseases; buy approved or certified stock plants (see Approved stock, Certified stock).

Wallflower *Cheiranthus cheiri* varieties. Hardy herbaceous perennial. Height: small to medium. Flowering time: spring. Flower colours: cream, orange, pink, purple, reddish-brown, white or yellow. Use: spring bedding, outdoor containers, window boxes; usually treated as a biennial; may be seen as a perennial on walls. It is not generally appreciated that Wallflower are botanically similar to Brassicas, and are subject to the same troubles.

Pests and Diseases. 1 Black mould, caused by the fungi *Alternaria brassicola* and *Alternaria cheiranthi*. May attack young and established plants. **Symptoms:** roundish grey areas on the leaves. **Control:** seldom damaging to any great extent, but spray with a copper fungicide. **2** Club root (see Brassicas 15 Club root or Finger and toe). **3** Damping off (see Brassicas 17 Damping off or wire stem). **4** Downy mildew (see Brassicas 20 Downy mildew). **5** Grey mould (see Brassicas 23 Grey mould). **6** Mildew (see Brassicas 26 Mildew). **7** Virus, Mosaic virus (see Brassicas 35b Virus). Note that as well as mottling on the leaves, there may be colour breaking or flower breaking, that is, a red-flowered variety, for example, may show a proportion of red/yellow flowers. **8** White blister (see Brassicas 38 White blister).

Waterlily *Nymphea alba* Common white waterlily and *Nymphea* varieties. Hardy, aquatic, herbaceous perennials, with fleshy or tuberous roots. Greenhouse. Height: some dwarf varieties, widespreading over water, flowers floating or held above water. Flowering time: summer to autumn. Flower colours: cream, blue, lilac, pink, purple, red, white, yellow or parti-coloured. Fragrant. Leaves: large, decorative, shaded with a contrasting colour. Use: greenhouse pools or containers, large varieties outside in lakes and pools, dwarf varieties in small pools or containers.

Pests and Diseases. 1 Caddis fly, May fly *Limnephilus marmoratus*. May attack established plants. **Symptoms:** damage to leaves caused by attached larvae. **Control:** hand-pick if possible; clear up rotting material below water. Fish will help to reduce numbers. **2** Caterpillars of brown china mark moth *Nymphula nympheata*. May attack established plants. **Symptoms:** pale caterpillars living between two oval sections of leaf, tied together with silk, from early July to late August; ragged holes on lily pads (leaves) and rotting leaf tissue caused by caterpillars browsing, sometimes under water. **Control:** on no account should insecticides be used to control this pest as they could be fatal to fish; hand-pick the caterpillars and this should be sufficient to reduce damage. **3** Leaf spot caused by the fungus *Ovularia nymphearum* (see Leaf spot). May attack established plants. **Symptoms:**

unsightly marking and spotting of leaves. **Control:** remove badly affected leaves and even parts of leaves. **4** Stem rot, caused by a *Phytophthora* species of fungus. May attack established plants. **Symptoms:** blackening and rotting of stems. This is often associated with leaf spot (see 3 above Leaf spot). **Control:** remove and destroy affected plants, but if the disease persists, it may be necessary to use a fungicide. Always seek advice before doing this, from the local horticultural adviser or other authority. **5** Water lily aphis *Rhopalosiphum nympheae*. May attack established plants. **Symptoms:** distortion of leaves and stem; discoloured flowers. These pests can be damaging. **Control:** look for the first signs of aphids in early summer and spray at once with Nicotine. **6** Water lily beetle *Galerucella nympheae*. May attack established plants. **Symptoms:** dark brown beetles and larvae feeding on leaves, causing damage which is followed by rotting. **Control:** spray with insecticide suitable for water plants. Derris must *never* be used if fish are present. Fish feed on this pest, and will clear infested leaves if these are fastened by hoops or other means below the water for a day or two.

Weather When a plant shows unhealthy symptoms, rather than assume that a particular pest or disease is involved, look first of all for a cultural or physiological reason. Plants are often very badly affected by excesses of cold, heat, wind, drought and rain. Trees and shrubs, because of their size, are particularly prone to weather damage. See also Drought, Frost and cold, Rainfall, Sun and Wind for further information.

Weeds Unwanted plants in the garden are called weeds. They may be annual, perennial, biennial, herbaceous (soft stemmed and dying down in Winter) or woody. Weed roots can be surface or deep rooted, stoloniferous or creeping. Weeds compete with crops and plants and are unsightly. Weeds can be controlled mechanically by hoeing, or chemically by spraying or local application. Chemical weed-killers are called herbicides and may act through the soil via the plant roots, or be absorbed by the leaves. They may be contact or translocated (distributed through the plant). Herbicides may be total, killing all growth; selective, killing the weeds and leaving the plants unharmed; pre or post emergence, before or after the 'crop' plant develops; short term or residual. Soil sterilants (see Soil sterilization) are in effect herbicides. Herbicides can be specific to weeds and crops and it is advisable to read the reference

1 *Bindweed* 2 *Mouse-ear Chickweed* 3 *Bellbine* 4 *Chickweed*
5 *Coltsfoot* 6 *Sowthistle* 7 *Ground Elder* 8 *Onion Couchgrass*
9 *Couchgrass* 10 *Curled Dock* 11 *Horsetail*

publications of the chemical companies. Much useful information is contained in the *Directory of Garden Chemicals* (see Appendix p. 245–54) published by The British Agrochemicals Association. Remember that it is essential to apply herbicides as directed, avoid crop contamination or drift, and wash out sprayers.

Weevils   A large group of insects similar to beetles, but with differently-shaped heads, which are formed like snouts. The fat,

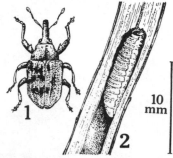

whitish larvae, unlike beetle larvae, are legless, and rather curved, with large brown heads. Weevil larvae tend to feed inside buds, flowers, fruits, shoots, stems, roots, corms and tubers. The habits of adult weevils vary according to species, some feeding by night and sledom seen, others feeding by day. Often pieces are bitten out of leaves.

Stem borer 1 Weevil   2 Larva

Control: spray with Carbaryl, gamma-HCH or Fenitrothion. In some cases, a gamma-HCH seed-dressing may be very effective, followed by spraying with gamma-HCH.

White currant   (see Red and white currants)

Whitefly   Various insects are known as whitefly. They are related to aphids and scale insects (see Aphis, Scale insects). Different species of whitefly may attack a wide range of plants, in the greenhouse or in the open. Adult insects look like minute moths, but are covered with a white, waxy substance. They can be seen in large numbers on young plants and on the growing tips of old plants. When disturbed, they rise in a white cloud. Like aphids, they excrete honeydew which attracts sooty moulds (see Honeydew). Sooty moulds block light from the leaves and disfigure them. Whitefly attacks are worst in late summer. In severe cases, they can spoil or even destroy plants.

Control: use HCH smokes, or gamma-HCH sprays weekly, or sprays of Bioresmethrin, Malathion, Pirimiphos-methyl, or Resmethrin twice weekly, or Dichlorvos-impregnated blocks; in the greenhouse, use parasitic wasps as resistance to pesticides is common (see Biological control).

Willow   *Salix alba* white willow, *Salix chryscoma* weeping willow, *Salix caprea* goat willow, *Salix herbacea* dwarf willow and other

Salix species and varieties. Hardy deciduous trees and shrubs. Height: trees, small to medium; shrubs, dwarf and prostrate to medium. Flowering time: late winter to early spring. Flowers: decorative catkins. Flower colours: grey, pink or purple, turning white or yellow. In some sorts, stems may be black, red or yellow. Use: planted singly, in groups or in rows, ground cover, rock gardens and especially suitable for water-side planting.

Pests and Diseases. 1 Anthracnose, caused by the fungus *Marssonina salicicola*. May attack young and established plants. **Symptoms:** brown spots on the leaves and dark marks on the stems. **Control:** cut out diseased parts; spray with fungicide to protect remaining healthy parts. **2** Beanseed gall sawfly *Pontania capreae*. May attack young and established plants. **Symptoms:** about midsummer, galls of the same shape, size and colour as runner bean seeds, in the middle of leaves; a single white sawfly caterpillar may be found in each gall. **Control:** hand-pick to remove this minor disfigurement. **3** Black blister, caused by the fungus *Cryptomyces maximus*. May attack established plants. **Symptoms:** black blisters along the branches. **Control:** cut out infected branches. **4** Black canker, caused by the fungus *Physalospora miyabeana*. May attack established plants. **Symptoms:** disease at tips of leaves, spreading down in a wedge shape, and causing tips to bend sharply; leaves may be killed; cankers develop on stems. **Control:** spray with Bordeaux mixture in the early part of the year. **5** Blue and brassy willow beetles *Phyllodecta* species. May attack young and established plants. **Symptoms:** about mid-summer, small, metallic-coloured beetles stripping the soft parts of leaves and leaving only a skeleton of veins; similar damage caused by the larvae which are small and black. In severe attacks, trees may take on a scorched appearance. **Control:** shake branches of the affected tree over an inverted umbrella, and collect the beetles, which can then be destroyed. Damage is never so severe as to require the killing of the tree. The above control will prevent a certain amount of disfigurement. **6** Galls (see 2 above Beanseed gall sawfly). **7** Scale, Willow scale *Chionaspis salicis* (see Scale insects). May attack young and established plants. **Symptoms:** white scales encrusting the young stems, causing stunting. This condition is noticeable in summer. **Control:** spray with Diazinon or Malathion in spring.

Wilt   Plants may wilt because they are short of water and except in the case of extreme water loss, they will recover when water is supplied again. But wilting may be due to the presence of various diseases, usually caused by certain fungi attacking the roots, clogging the conducting tissue in the stems and sending material made by the fungus into the plant sap. There are many examples of wilt diseases in cultivated plants. Chemicals may be used in controlling them. For specific control, see under plant affected.

Wind Browning of leaf margins is a typical symptom of wind damage. Ornamental acers and other soft leaved trees or shrubs are especially prone to attack. Persistent wind damage shapes trees badly and indeed many species cannot become established in exposed areas unless shelter of some sort is provided.

Winter wash Pesticides (insecticides) used chiefly on fruit trees during their dormant period to control overwintering pests. If used when growth has commenced, damage to new growth is likely, as winter washes (mainly tar oil) are damaging to growth. Winter washes are also useful for a wide range of woody trees or shrubs. They remove moss and algae by chemical burning.

Wireworms Larvae of the click beetles, which include *Agriotes* and *Athous* species, are called wireworms. They are 6-legged, thin

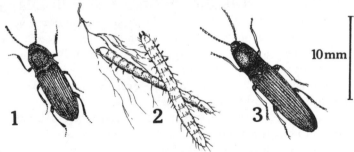

1 Agriotes and 3 Athous species of click beetle   2 Larvae (wireworms)

and wiry, cream-coloured when young, darkening to brown as they grow. Wireworms are universally present in grass and turf soil, where the young larvae begin by feeding on grass roots and other organic matter, becoming beetles after several years. As they grow, and if, for instance, grass land is dug up for planting, older larvae will attack a wide range of plants. They feed on stems at ground level, roots, bulbs, corms, tubers and rhizomes, doing considerable damage.

Control: in the open, control is difficult; in dug up grassland, potatoes are a good cleaning crop which should be lifted early and the wireworm-riddled tubers destroyed; mustard sown as a crop in July and dug in when it is about 7.5 cm (3 in) is also useful; use HCH and Naphthalene wireworm dusts (if available); in the greenhouse, sterilize border soil and composts (see Soil sterilization).

Woodlice   These pests are also known as pillbugs or slaters. There are several different species, including the common woodlouse

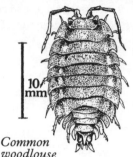

10/
mm

*Common
woodlouse*

Armadillidium vulgare, the garden woodlouse *Oniscus asellus* and the grey garden woodlouse *Porcellio scaber*. Woodlice feed on dead and dying organic matter of any sort, but they will also attack plants in the open and in the greenhouse by gnawing the stems, eating holes in the leaves or eating the tips of pot plant roots. They feed by night and hide by day in convenient places. They may do considerable damage if not checked.

Control: use traps, which can be made of hollowed-out potato tubers, orange or grapefruit peel halves or small inverted flower pots filled with some old organic matter, which should be put down at night and cleared out every morning; use some of the proprietary poison baits or make poison bait from 114 g (4 oz) of Paris Green, mixed with 3 kg (7 lb) bran, with or without 2 litres (½ gal) water; sterilize composts or soil (see Soil sterilization).

Z

Zantedeschia   (see Arum lily)

Zinnia   *Zinnia elegans* varieties. Half-hardy annual. Height: small to medium. Flowering time: summer. Flower colours: cream, green, lilac, orange, pink, red, yellow, or parti-coloured. Use: annual borders, summer bedding, outside containers, window boxes, greenhouse pot plants.

Zinnia 1 Grey mould

Pests and Diseases. 1 Grey mould, caused by the fungus *Botrytis cinerea* (see Botrytis). May attack established plants. **Symptoms:** grey mould on foliage. May be troublesome in wet weather. **Control:** in the open, remove badly affected plants and improve growing conditions; in the greenhouse, avoid overcrowding and excessive humidity by spacing out plants and reducing humidity. **2** Seedling blight, caused by the fungus *Alternaria zinniae*. May attack seedlings. **Symptoms:** dark brown cankers on the stems and somewhat angular reddish-brown spots on the leaves. **Control:** spray seedlings with Zineb or a proprietary copper fungicide. The fungus is seed-borne, and a seed dressing with Captan will reduce infection.

Glossary of Terms

Annual A plant raised from seed which can, if hardy, be sown in the autumn, or if half hardy or tender, in the spring. An annual has only one generation, producing flowers. The seed produced may fall to the ground and germinate on its own at the right time, but more usually, and necessarily with tender annuals, the seed is collected for sowing again, to produce another generation in the following year.
Example: Common or Scotch marigold.

Biennial A plant raised from seed, sown one year to flower the next, when it will produce seed and die.
Example: Hollyhock.

Blanching Depriving plant leaves of light so that they become white or yellowish.

Bleaching The lightening or whitening of leaf colour due to nutritional or physiological upset, or pest or disease attack.

Bract The upper portion of a plant exposed to light, usually but not always leaves.

Breaking The change in form or colour of a plant, caused genetically or by pest or virus disease.

Bud burst Period when fruit and growth of buds on deciduous trees is evident.

Bulb Swollen underground leaves which protect the bud and serve as a food supply for the embryo plant.
Examples: Daffodil, Hyacinth.

Catalyst An element or chemical which promotes chemical change but remains unchanged itself. Chlorophyll is a catalyst in the process of photosynthesis.

Chlorophyll The green pigment contained in green plants which acts as a catalyst in the processes of food manufacture and photosynthesis.

Glossary

Corm   The resting form of some plants, consisting of part of a stem at ground level, usually swollen and containing a reserve of food. Examples: Crocus, Gladiolus.

Crotch   The area of branch development on woody plants.

Deciduous   Shedding leaves annually. Examples: Oak, Rose.

Dieback   The death of a portion of a stem or branch usually caused by extreme cold, or by bad pruning.

Drupelets   Individual globules making fruits such as blackberry and raspberry.

Evergreen   Having green leaves for at least a year, sometimes longer. Examples: Silver fir, Laurel.

Fasciation   Abnormal growth of leaves and stem, stimulated by disease or excess hormones. Stems may be flattened and leaves twisted.

Frass   The desiccated tissue and exudation left behind by pests eating internally in plant tissue.

Fruiting body   The reproductive portion of certain fungi containing spores, sometimes appearing above the ground.

Gall   An abnormal rounded growth developing on a plant, caused by certain pests and diseases.

Germination   The process of emergence of a young plant from a seed. Circumstances favourable to germination vary according to type of seed. Sufficient supplies of air and moisture at the right temperature are usually required. For healthy development, it is important to ascertain the right conditions for the particular type of seed.

Growing medium   The material in which a plant is grown in pots or containers. It can be soil, peat, sand or any one of a number of materials with suitable physical qualities.

Herbaceous plant Plant without a woody stem. Herbaceous plants may be annuals, biennials or perennials.

Laterals Branches or shoots arising from main stems or trunks of bushes, trees or plants.

Medium (see Growing medium)

Mulch A layer of material, usually but not always organic, spread on the soil surface to conserve moisture, supply nutrients, or control weeds.

Node The area of a plant where a leaf is attached to a stem.

Perennial A plant which continues to live for many years. Herbaceous perennials lose their leaves and die down, more or less to ground level in winter in temperate zones. Woody perennials such as shrubs and trees which are deciduous, lose their leaves in winter in temperate zones, but in milder areas may be semi-evergreen and lose only a portion of their leaves. Their stems or branches remain more or less permanent, according to whether the tree or shrub in question is hardy enough to stand up to cold weather. Evergreen perennials remain in leaf continuously, although a percentage of the leaves may fall, to be replaced in time with new leaves.
Example: Privet.

Photosynthesis The synthesis of simple carbohydrates from water and carbon dioxide, light being the energy source. Photosynthesis takes place mainly on the leaves, so if leaves are damaged, the vital process will be hindered.

Plant, Parts of Root: part of a plant underground, which absorbs water and nutrients. Stem: part of a plant above ground, which carries water from roots to leaves. Leaf: part of a plant which is essential to the processes of breathing, photosynthesis and transpiration. Flower: the regenerative part of a plant where the seed for the next generation is formed.

Rhizome   A horizontal creeping underground portion of a plant stem, usually thick and swollen. Plants can be reproduced from part of a rhizome.

Rhizomorph   Long vegetative strand of fungal growth, enabling the fungus to grow considerable distances through soil in search of food.

Scorch   Injury to plant foliage caused by excess of dry heat in greenhouse and by wind or sun in the garden.

Shrub   Woody branching plant with no main trunk.

Spore   The reproductive body of a fungus, usually a separate cell.

Sub-shrub   Partly herbaceous and partly woody plant.

Tuber   A swollen underground stem portion, usually roundish, with buds or eyes from which new plants or other tubers may grow. Example: Potato.

Appendix of Garden Chemicals

This appendix lists the majority of the recommended chemicals and proprietary brands readily available to amateur gardeners from retail horticultural outlets. Other recommended chemicals, not listed in the appendix, are available either through retail outlets or agricultural/horticultural merchants, sometimes in larger quantities only. This is especially the case with those chemicals which are described in the text as being commercially available and which are included in the *Agricultural Chemicals Approval Scheme* booklet, produced annually by the United Kingdom Ministry of Agriculture, Fisheries and Food, available from divisional offices of the Ministry of Agriculture—or in Scotland, from the Department of Agriculture for Scotland.

The following information is taken from the *Directory of Garden Chemicals* issued by the British Agrochemicals Association. Copies of the Directory are available from The British Agrochemicals Association, Alembic House, 93 Albert Embankment, London SE1.

ANA and Captan *Pesticide*
1 Murphy Hormone Rooting Powder, a powder for rooting cuttings without rotting. *Murphy Chemical Ltd.*
2 Strike, a powder rooting hormone. *May and Baker Ltd.*

Anthraquinone *Pesticide*
1 Morkit, a bird repellent. *May and Baker Ltd.*

Benomyl *Fungicide*
1 Benlate, a wettable powder for the control of a wide range of diseases in the garden and greenhouse. *Pan Britannica Industries Ltd.*

Bioresmethrin *Insecticide*
1 Cooper Garden Spray, a spray for the control of blackfly, greenfly, caterpillars, sawfly, capsid bugs, raspberry beetles, etc. on edible crops. *Wellcome Foundation Ltd.*
2 Fisons Combat Whitefly Insecticide, a liquid and an aerosol to kill whitefly particularly in the greenhouse. *Fisons Ltd.*

Bioresmethrin and Malathion *Insecticide*
1 Fisons Combat Vegetable Insecticide, a liquid and an aerosol for control of a wide range of fruit and vegetable pests. *Fisons Ltd.*

Borax *Insecticide*
1 Nippon Ant Destroyer, a liquid to control ants. *Synchemicals Ltd.*

Bordeaux mixture (see Copper Compound)

Bromophos *Insecticide*
1 Bromophos, a coarse powder for the control of soil pests. *Pan Britannica Industries Ltd.*

Calomel (see Mercurous Chloride) *Fungicide*

Calomel (see Mercurous Chloride) *Insecticide*

Captan *Fungicide*
1 Murphy Orthocide Captan Dust, a dust (puffer pack) for the control of damping off, grey mould, bloom damping. *Murphy Chemical Ltd.*
2 Murphy Orthocide Captan *Fungicide*, a wettable powder (sachets) for the control of apple scab, black spot, Botrytis, bulb rots, etc. *Murphy Chemical Ltd.*

Captan and Lindane *Fungicide and Insecticide*
1 Murphy Combined Seed Dressing, a powder (sachet) for the control of pests and diseases on seeds and seedlings. *Murphy Chemical Ltd.*

Carbaryl *Insecticide*
1 Boots Garden Insect Powder, a dust for the control of caterpillars etc. *The Boots Company Ltd.*
2 Murphy Sevin Dust, a dust (puffer pack) for the control of caterpillars, weevils, beetles, etc. *Murphy Chemical Ltd.*
3 Murphy Wasp Destroyer, a powder for the control of wasps (destruction of nests). *Murphy Chemical Ltd.*

Carbaryl/Quintozene *Insecticide*
1 Autumn Toplawn. *Pan Britannica Industries Ltd.*

Carbaryl and Quintozene *Pesticide*
1 Autumn Toplawn, a powder for the control of worms and diseases in lawns. *Pan Britannica Industries Ltd.*

Appendix

Carbendazim and Maneb *Fungicide*
1 Fisons Combat Rose Fungicide, a powder (sachets) for the control of mildew and blackspot on roses. *Fisons Ltd.*

Cheshunt Compound (see Copper Sulphate and Ammonium Carbonate)

Chlordane *Insecticide*
1 Chlordane 25. *Synchemicals Ltd.*
2 Murphy Ant Killer Liquid, a liquid for the control of ants. *Murphy Chemical Ltd.*
3 Murphy Ant Killer Powder, a powder (puffer pack) for the control of ants. *Murphy Chemical Ltd.*
4 Murphy Chlordane Wormkiller, a liquid for the control of worms. *Murphy Chemical Ltd.*
5 Nippon Ant Powder, a powder for the control of ants. *Synchemicals Ltd.*

Chlordane/ Piperonyl Butoxide/ Pyrethrins *Insecticide*
1 Cooper Ant Killer Aerosol, a spray for the control of ants. Also kills cockroaches, beetles and silverfish. *Wellcome Foundation Ltd.*
2 Nippon Aerosol Spray, a liquid aerosol for the control of ants, wasps and crawling insects. *Synchemicals Ltd.*

Chlorpyrifos *Insecticide*
1 Murphy Soil Pest Killer, granules (shaker pack) for the control of soil pests in vegetable crops. *Murphy Chemical Ltd.*

Copper Compound *Fungicide*
1 Murphy Bordeaux Powder, a wettable powder for use as a spray or dust for the control of blight, leaf spot, canker and rose diseases. *Murphy Chemical Ltd.*
2 Murphy Liquid Copper Fungicide, a liquid for the control of damping off of seedlings, leaf mould, blight, mildew, rust, etc. *Murphy Chemical Ltd.*
3 Murphy Rose Mildew Specific, a liquid for the control of blackspot, mildew, rust on roses. *Murphy Chemical Ltd.*

Copper Sulphate and Ammonium Carbonate *Fungicide*
1 Cheshunt Compound, a soluble powder to prevent damping off in seedlings. *Pan Britannica Industries Ltd.*

Coumatetralyl *Pesticide*
1 Racumin Mouse Bait, a powder bait for mice. *Pan Britannica Industries Ltd.*
2 Racumin Rat Bait, a powder bait for rats. *Pan Britannica Industries Ltd.*

Derris (see Rotenone) *Insecticide*

Diazinon *Insecticide*
1 Fisons Combat Soil Insecticide, granules for the control of a wide range of common soil pests. *Fisons Ltd.*

Dichlofluanid *Fungicide*
1 Elvaron, a powder for the control of blackspot on roses. *May and Baker Ltd.*

Dichlorophen *Fungicide*
1 Murphy Super Moss Killer and Lawn Fungicide, a liquid for the control of moss, liverworts lichen, algae on turf, paths, fences, etc. and lawn disease control. *Murphy Chemical Ltd.*

Dichlorvos *Insecticide*
1 Vapona Flykiller, in the form of solid impregnated PVC for the control of flies, under glass only.

Dicofol / Dinocap / Fenitrothion, Maneb / Pyrethrum *Fungicide and Insecticide*
1 Murphy Combined Pest and Disease Spray, a spray (aerosol) for the control of pests and diseases on a wide range of crops. *Murphy Chemical Ltd.*

Dimethoate *Insecticide*
1 Boots Systemic Greenfly Killer, a liquid for the control of blackfly and greenfly on roses, flowers and vegetables. *The Boots Company Ltd.*
2 Murphy Systemic Insecticide, a liquid for the systemic control of greenfly, whitefly, red spider, sawfly, mealybug, scale insects, etc. *Murphy Chemical Ltd.*

Dinocap *Fungicide*
1 Murphy Dinocap Dust, a dust (puffer pack) to control powdery mildew. *Murphy Chemical Ltd.*
2 Murphy Dinocap Mildew Fungicide, a wettable powder (sachet) for the control of powdery mildew. *Murphy Chemical Ltd.*
3 Murfume Dinocap Smoke (Cone), a fumigant for the control of powdery mildew under glass. *Murphy Chemical Ltd.*
4 Toprose Mildew Spray, a liquid for the control of mildew on roses. *Pan Britannica Industries Ltd.*
5 XLALL Mildew Wash, a liquid for the control of powdery mildew on roses, chrysanthemums, etc. *Synchemicals Ltd.*

Dinocap and Folpet *Fungicide*
1 Murphy Rose Fungicide, a wet-table powder (sachets) for the control of blackspot and mildew on roses. *Murphy Chemical Ltd.*

DNOC and Petroleum Oil *Insecticide*
1 Murphy Ovamort Special, a liquid winter wash for fruit trees, for red spider, aphids, capsids, etc. *Murphy Chemical Ltd.*

Fenitrothion *Insecticide*
1 Murphy Fentro, a liquid for the control of greenfly, sawfly, caterpillars, capsids, beetles, etc. *Murphy Chemical Ltd.*
2 PBI Fenitrothion, a liquid for the control of greenfly, blackfly, capsids, and caterpillars in fruit and vegetables; codling moth and sawfly (maggots in fruit) in apples and pears, raspberry beetle. *Pan Britannica Industries Ltd.*

Formaldehyde *Fungicide*
1 Murphy Formaldehyde Soil Sterilizer, a liquid soil sterilant for use in greenhouse and outdoors. *Murphy Chemical Ltd.*

Formalin (see Formaldehyde)

Formothion *Insecticide*
1 Topgard Systemic Liquid, a liquid for the control of greenfly, blackfly, red spider and small caterpillars in the garden. *Pan Britannica Industries Ltd.*
2 Toprose Systemic Spray, a liquid for the control of greenfly, red spider, leafhopper and small caterpillars on roses. *Pan Britannica Industries Ltd.*

gamma-HCH *Insecticide*
1 Fumite Lindane Smoke (Pellets), fumigant to control greenhouse pests. *May and Baker Ltd.*

Appendix

2 ICI Antkiller, a dust for the control of ants, earwigs, woodlice and wasps. *ICI Plant Protection Division.*
3 ICI Root Fly and Wireworm Dust, a dust for the control of soil pests. *ICI Plant Protection Division.*
4 Murphy Gamma-BHC Dust, a dust for the control of rootfly, wireworm, beetles leatherjackets, etc. *Murphy Chemical Ltd.*
5 Murfume Lindane Smoke (Pellets), smoke pellets for the control of greenfly, whitefly, capsids, leaf miner, etc. *Murphy Chemical Ltd.*
6 Murphy Lindex Garden Spray, a liquid for the control of greenfly, earwig, thrips, caterpillars, rootfly, etc. *Murphy Chemical Ltd.*

gamma-HCH/Dimethoate/ Malathion *Insecticide*
1 Super Kil, a liquid systemic insecticide which controls a broad spectrum of most common garden and greenhouse pests. *Fisons Ltd.*

gamma-HCH/Pyrethrins/ Piperonyl Butoxide *Insecticide*
1 Boots Garden Insect Killer, an aerosol spray for the control of blackfly and greenfly. *The Boots Company Ltd.*

gamma-HCH/Rotenone/Thiram *Insecticide and Pesticide*
1 Hexyl Plus, a combined liquid insecticide and fungicide for general garden use. *Pan Britannica Industries Ltd.*

gamma-HCH/Tecnazene *Insecticide*
1 Fumite Smoke Cones, a smoke for the control of greenhouse pests. *May and Baker Ltd.*

gamma-HCH and Malathion *Insecticide*
1 New Kil, a spray contact insecticide designed for control of small outbreaks of garden pests. *Fisons Ltd.*

gamma-HCH and Menazon *Insecticide*
1 Abol-X, a liquid for the systemic control of greenfly and blackfly. *ICI Plant Protection Division.*

HCH *Insecticide* (see also gamma-HCH) 1 Murfume BHC Smoke (Cone), a smoke fumigant for the control of greenhouse pests. *Murphy Chemical Ltd.*
2 Fumite HCH Smoke (Cones), a smoke for the control of greenhouse pests. *May and Baker Ltd.*

Lindane/Malathion/ Dimethoate *Insecticide*
1 Fisons Combat Garden Insecticide, a liquid to kill greenfly and all major pests. *Fisons Ltd.*

Lindane and Malathion *Insecticide*
1 Fisons Combat Garden Insecticide Aerosol, an aerosol to kill all major pests. *Fisons Ltd.*

Malathion *Insecticide*
1 Boots Greenfly Killer, a liquid for the control of blackfly and greenfly. *The Boots Company.*
2 Malathion Greenfly Killer, a liquid for the control of greenfly, blackfly, whitefly, red spider, thrips, leaf miner, woolly aphids, mealybugs and scale insects. *Pan Britannica Industries Ltd.*
3 Murphy Greenhouse Aerosol, a spray for the control of greenfly, whitefly, red spider, mealybug, etc.

Murphy Chemical Ltd.
4 Murphy Liquid Malathion, a liquid for the control of greenfly, whitefly, red spider, mealybug, scale insects, etc. *Murphy Chemical Ltd.*
5 Murphy Malathion Dust, a dust for the control of greenfly, thrips, whitefly, etc. *Murphy Chemical Ltd.*

Malathion and Dimethoate
Insecticide
1 Vitax Greenfly/Blackfly Spray; a liquid for the control of greenfly, blackfly, whitefly, red spider, raspberry beetle, leaf miner, thrips, scale insects and leaf hoppers. *Steetley Chemicals Ltd.*

Mercurous Chloride (Calomel)
Fungicide
1 PBI Calomel Dust, a powder to control moss in lawns and clubroot in brassicas. *Pan Britannica Industries Ltd.*
2 Cyclosan, a powder for the control of club root and white rot. *May and Baker Ltd.*
3 ICI Club Root Control, a dust for the control of club root on brassicas. *ICI Plant Protection Division.*
4 Murphy Calomel Dust, a dust (puffer pack) for the control of club root and onion white rot. *Murphy Chemical Ltd.*

Mercurous Chloride *Herbicide*
1 Calomel Dust, a powder to control moss in lawns and clubroot in brassicas. *Pan Britannica Industries Ltd.*
2 Cyclosan 4% Calomel Dust. *May and Baker Ltd.*
3 Mos-Tox, a powder for moss control in lawns. *May and Baker Ltd.*
4 M-C, a powder for moss control in lawns. *Synchemicals Ltd.*

Mercurous Chloride (Calomel)
Insecticide
1 Boots Calomel Dust, a powder for the control of club root and cabbage root maggot. *The Boots Company Ltd.*
2 Calomel Dust. *Pan Britannica Industries Ltd.*
3 Cyclosan, a powder for the control of club root, cabbage root fly and onion fly. *May and Baker Ltd.*
4 Murphy Calomel Dust, a powder (puffer pack) for the control of carrot and onion fly. *Murphy Chemical Ltd.*

Mercury as Phenyl, Mercury Acetate *Fungicide*
1 Verdasan, a water dispersible powder for the control of fungal diseases on turf. Verdasan is a listed poison. Retail sales to private individuals may only be made by chemists. 'Listed sellers' may sell to bona fide sports clubs on receipt of a signed order. *ICI Plant Protection Division.*

Metaldehyde *Pesticide*
1 Boots Slug Destroyer, pellets for the control of slugs and snails. *The Boots Company Ltd.*
2 Fisons Combat Slug Pellets, pellets to control slugs and snails. *Fisons Ltd.*
3 ICI Slug Pellets, mini pellets for the control of slugs and snails. *ICI Plant Protection Division.*
4 Murphy Slugit Liquid, a liquid for the control of slugs and snails. *Murphy Chemical Ltd.*
5 Murphy Slugit Pellets, pellets for the control of slugs and snails. *Murphy Chemical Ltd.*
6 Slug Mini-Pellets, pellets for the control of slugs and snails. *Pan Britannica Industries Ltd.*

Appendix

Methiocarb *Pesticide*
1 Draza, mini pellets for the control of slugs and snails. *May and Baker Ltd.*

Nicotine *Insecticide*
1 XLALL Insecticide, a liquid for the control of aphids, for use on vegetables, flowers, etc. *Synchemicals Ltd.*

Oxycarboxin *Fungicide*
1 Plantvax 75, a dispersible powder for the control of rust on ornamentals such as roses, chrysanthemums, carnations, pelargoniums. *ICI Plant Protection Division.*

Oxydemeton-methyl *Insecticide*
1 Greenfly Gun, an aerosol for the control of greenfly and blackfly on crops. *May and Baker Ltd.*

Pirimicarb *Insecticide*
1 Rapid Greenfly Killer, a liquid and an aerosol spray for the quick control of greenfly and blackfly. *ICI Plant Protection Division.*
2 Rapid Aerosol. *ICI Plant Protection Division.*

Pirimphos-methyl *Insecticide*
1 ICI Antkiller, a dust for the control of ants, earwigs, woodlice and wasps. *ICI Plant Protection Division.*
2 Sybol 2, a liquid for the control of most garden and greenhouse pests including whitefly and red spider. *ICI Plant Protection Division.*
3 Sybol 2 Dust, a dust for the control of soil and vegetable pests. *ICI Plant Protection Division.*

Pirimphos-methyl and Synergised Pyrethrins *Insecticide*

1 Sybol 2 Aerosol, an aerosol spray for the control of most garden and greenhouse insect pests including whitefly and red spider. *ICI Plant Protection Division.*
2 Kerispray, an aerosol spray for general insect control (including whitefly and red spider) on houseplants. *ICI Plant Protection Division.*
3 Waspend, an aerosol spray for the control of flying and crawling insects in the home. *ICI Plant Protection Division.*

Pyrethrins and Piperonyl Butoxide *Insecticide*
1 Coopers Insect Powder, a dusting powder for the control of garden and greenhouse pests, e.g. greenfly, blackfly, aphids, thrips, flea beetles, caterpillars, earwigs, wasps, etc. *Wellcome Foundation Ltd.*

Pyrethrum *Insecticide*
1 Anti-Ant Powder, a powder for the control of ants indoors and out of doors. *Pan Britannica Industries Ltd.*
2 Plant Pest Killer, an aerosol spray for the quick control of aphids, etc. and for use on vegetables, flowers, etc. *Synchemicals Ltd.*

Pyrethrum and Resmethrin *Insecticide*
1 Bio Sprayday, a liquid for the control of whitefly, greenfly, blackfly, outdoor ants, leafhopper and thrips in the garden, greenhouse and on houseplants. *Pan Britannica Industries Ltd.*

Quintozene/Carbaryl *Fungicide*
1 Autumn Toplawn, a powder for the control of worms and diseases in lawns. *Pan Britannica Industries Ltd.*

Resmethrin/Malathion/ Trichlorphon *Insecticide*
1 Crop Saver, a liquid for the control of insects on vegetable crops. *Pan Britannica Industries Ltd.*

Rotenone (Derris) *Insecticide*
1 Abol Derris Dust, a dust for general insect control, particularly on soft fruit. *ICI Plant Protection Division.*
2 Boots Derris Dust, a powder used as a general insecticide. *The Boots Company Ltd.*
3 Liquid Derris, a liquid for the control of greenfly, blackfly, thrips, red spider, small caterpillars, sawfly and raspberry beetle grubs. *Pan Britannica Industries Ltd.*
4 Murphy Derris Dust, a powder (puffer pack) for the control of greenfly, caterpillars, beetles, etc. *Murphy Chemical Ltd.*
5 Murphy Derris Liquid, a liquid for the control of greenfly, caterpillars, beetles, etc. *Murphy Chemical Ltd.*

Rotenone/Sulphur/Zineb *Insecticide*
1 Murphy Combined Pest and Disease Dust, a powder (puffer pack) for the control of pests and disease on all crops. *Murphy Chemical Ltd.*

Sodium Chlorate *Herbicide*
1 Boots Sodium Chlorate, crystals for the control of deep rooted weeds, nettles and grass. *The Boots Company Ltd.*

Sulphur *Fungicide*
1 Murphy Lime Sulphur, a liquid for the control of big bud mite, peach leaf curl etc. *Murphy Chemical Ltd.*

Sulphur *Pesticide*
1 Murphy Mole Smoke, a smoke canister for the control of moles, rats and mice outdoors. *Murphy Chemical Ltd.*

Sulphur/Zineb/Rotenone *Fungicide and Pesticide*
1 Murphy Combined Pest and Disease Dust, a powder (puffer pack) for the control of pests and diseases on all crops. *Murphy Chemical Ltd.*

Tar Oil *Herbicide*
1 Murphy Moss and Speedwell Killer, a liquid for the control of moss and speedwell on turf, paths, etc. *Murphy Chemical Ltd.*

Tar Oil *Insecticide*
1 Murphy Mortegg, a liquid winter wash for roses, fruit trees etc. for pest control, moss, etc. *Murphy Chemical Ltd.*

Tecnazene *Fungicide*
1 Fumite Tecnazene, pellets for use as a smoke in the control of botrytis on tomatoes etc. *May and Baker Ltd.*

Thiophanate-methyl *Fungicide*
1 Murphy Systemic Fungicide, a wettable powder (sachets) for the control of blackspot, mildew, scab, leaf mould, turf diseases, etc. *Murphy Chemical Ltd.*

Thiram *Fungicide*
1 ICI General Garden Fungicide, a liquid for protection against blackspot, mildews, rusts and other fungal diseases. *ICI Plant Protection Division.*

Appendix

Thiram/gamma-HCH/ Rotenone *Fungicide and Insecticide*
1 Hexyl Plus, a combined liquid insecticide and fungicide for general garden use. *Pan Britannica Industries Ltd.*

Trichlorphon *Insecticide*
1 Dipterex, a wettable powder for the control of caterpillars on leaf vines and dipterous pests. *May and Baker Ltd.*
2 Tugon, granules for the control of ants. *May and Baker Ltd.*

Zineb *Fungicide*
1 Dithane, a wettable powder for the control of rust, blackspot, downy mildew, apple scab, peach leaf curl, potato and tomato blight. *Pan Britannica Industries Ltd.*